A Selection of Highlights from the History of the National Academy of Sciences, 1863–2005

Frederick Seitz

T0127889

UNIVERSITY PRESS OF AMERICA,® INC.
Lanham • *Boulder* • *New York* • *Toronto* • *Plymouth, UK*

Copyright © 2007 by
University Press of America,® Inc.
4501 Forbes Boulevard
Suite 200
Lanham, Maryland 20706
UPA Acquisitions Department (301) 459-3366

Estover Road
Plymouth PL6 7PY
United Kingdom

Library of Congress Control Number: 2006937513
ISBN-13: 978-0-7618-3586-8 (clothbound : alk. paper)
ISBN-10: 0-7618-3586-5 (clothbound : alk. paper)
ISBN-13: 978-0-7618-3587-5 (paperback : alk. paper)
ISBN-10: 0-7618-3587-3 (paperback : alk. paper)

∞™ The paper used in this publication meets the minimum
requirements of American National Standard for Information
Sciences—Permanence of Paper for Printed Library Materials,
ANSI Z39.48—1984

Contents

Terms of Presidents of the National Academy Of Sciences
   Featured in the Accompanying Historical Highlights   vii

Acknowledgements   ix

Founding   1
Alexander Dallas Bache   1
   The Smithsonian: *Joseph Henry*   3
Joseph Henry   6
William Barton Rogers   9
   John Wesley Powell   11
Othniel Charles Marsh   15
   Maxwell's Equations   18
   Henry Rowland   19
Oliver Wolcott Gibbs   20
   Muir-Pinchot Debate   22
Alexander Agassiz   22
   Revolution In Physics   24
Ira Remsen   25
   Control of Commercial Trusts   26
   Semicentennial Celebration   27
   Prediction of Neutron; Harkins-Chadwick   28
William Henry Welch   29
   Simon Flexner   30
   World War I: The National Research Council   32

Charles Doolittle Walcott 33
   Foundation Awards 35
   Versailles Treaty 38
Albert Abraham Michelson 39
   Edward Morley 41
Thomas Hunt Morgan 41
   The Great Depression 43
William Wallace Campbell 43
   James Lick 44
   George Davidson 46
   Benjamin Gould 48
Frank Rattray Lillie 48
Frank Barton Jewett 49
   Leslie R. Groves 52
   National Defense Research Committee 54
   James Conant 54
   Detlev W. Bronk 57
Alfred Newton Richards 58
Detlev Wulf Bronk 60
   The Battery Additive Crisis 61
   Mervin Kelly 61
   Texas Instruments 62
   The Trial of J. Robert Oppenheimer 63
   Harrison Brown 63
   The IGY: Lloyd Berkner 64
   Centennial Celebration 66
   Sputnik; The Creation of NASA 67
   The President's Scientific Advisory Committee 67
Frederick Seitz 69
   Harry Hess: Roger Revelle 70
   Vice President Hubert H. Humphrey 73
   The Enrico Fermi Laboratory 74
   Aldabra Islands 76
   Unidentified Flying Objects 76
   Tricentennial of The French Academy Of Sciences 78
   New Structures 79
   James A. Shannon 80
   Office of Technology Assessment 81
   A New Auditorium 82
   Hugh L. Dryden 83
   James E. Webb 84

Philip Handler   88
   Copernicus Celebration   90
   The Oklo "Natural" Nuclear Reactor   90
Frank Press   91
Bruce Michael Alberts   93
   Endowment   94
   Foreign Academies   94
   Housing The National Research Council   95
   An Emphasis on Education   95
   Bruce Alberts The Teacher   96
Ralph John Cicerone   96

Appendix A: Source of Photographs   99

Appendix B: Officers of the National Academy   103

Notes   107

Index   109

Terms of Presidents of the National Academy Of Sciences Featured in the Accompanying Historical Highlights

Alexander Dallas Bache	1863–1868
Joseph Henry	1868–1878
William Barton Rogers	1879–1882
Othniel Charles Marsh	1883–1895
Wolcott Gibbs	1895–1900
Alexander Agassiz	1901–1907
Ira Remsen	1907–1913
William Henry Welch	1913–1917
Charles Doolittle Walcott	1917–1923
Albert Abraham Michelson	1923–1927
Thomas Hunt Morgan	1927–1931
William Wallace Campbell	1931–1935
Frank Rattray Lillie	1935–1939
Frank Barton Jewett	1939–1947
Alfred Newton Richards	1947–1950
Detlev Wulf Bronk	1950–1962
Frederick Seitz	1962–1969
Philip Handler	1969–1981
Frank Press	1981–1993
Bruce Michael Alberts	1993–2005
Ralph John Cicerone	2005–

Acknowledgements

This informal history of the National Academy of Sciences could not have been produced without the very gracious and efficient professional help provided by Archivist Janice Goldblum of the Academy staff. My gratitude to her is unbounded.

In addition to emphasizing my great indebtedness to Janice Goldblum, it is a pleasure to recognize several other individuals who have been of help in the preparation of this specialized history of the Academy. John S. Coleman who served as executive officer of the Academy during part of my term in office read an early version of the manuscript and offered important suggestions. My long-time friend Dr. Robert N. Varney, a physicist and a fellow San Franciscan, read through large portions of the manuscript and, among many other things, provided me with very valuable information concerning Dr. Campbell's research at the Lick Observatory, as well as historical details in the development of the University of California at Berkeley where he completed his doctoral studies. Beyond this, he pointed out numerous errors and suggested additions that have been incorporated in the text. Dr. Marc Rothenberg, who is director-editor for the organization and publication of the collected papers of Joseph Henry at the Smithsonian Institution, read through the first half of the manuscript with great care and offered major improvements of the text out of his own highly expert knowledge. It has been a rare pleasure to follow the excellent work of Rothenberg and his staff. Ten of eleven volumes have been published so far. Rothenberg also generously provided Figure 51 from the Smithsonian Archives. Dr. Thomas Lassman, a professional historian of science then involved in a special project at the American Institute of Physics, scoured through the nearly completed manuscript and offered many improvements in the text that have been incorporated in it. Dr. Spencer Weart and Ms. Heather Lindsay of the staff of the American Institute of Physics generously provided photographs of Henry Rowland and Albert Einstein from the AIP archives (Figures 6 and 19). I am also indebted to Drs. Alexander Bearn and Frederick Wall, as well as Mrs. Lucie Marshall, for critical comments made after reading the text. Dr. Albert Adams, the current headmaster of Lick Wilmerding High School in San Francisco, reviewed some of the history of both the school and James Lick with me personally. He and his staff generously gave me the copy of the photograph of Lick that appears in this document. I am particularly indebted to Dr. and Mrs. Qinghong Yang for consistently valuable advice during the development of the entire manuscript as well as acquisition of the right to use photographs numbers 10 and 16 of Albert Einstein and Hideyo Noguchi. Rodney W. Nichols carried out a careful reading of the entire text and offered numerous valuable suggestions, which were adopted. Last but not least, I am, as usual, very grateful to Mrs. Florence Arwade, who manages my office, for her great help at all stages during the preparation of this manuscript.

I must here make special mention of the service I received from Dr. Carol Moberg, my colleague at The Rockefeller University who has a deep appreciation of the world of science. As a personal favor, she carried through a detailed reading of the accompanying text and made many useful suggestions.

We are particularly grateful to the staff of the George C. Marshall Institute in Washington, D.C. for crucial advice in acquiring a suitable publisher for this text. I am especially grateful to Jeff Kueter for his detailed help along the way and for forming close links with the staff of the publisher.

A Selection of Highlights from the History of the National Academy of Sciences,[1] 1863–2005

FOUNDING

The Academy's charter was introduced to Congress by the well-established Senator Henry Wilson from Massachusetts. It was passed by Congress and signed by President Lincoln in March of 1863, near the end of the second year following the start of the Civil War in April of 1861. The charter requested the members to serve the government on scientific and related technical problems without personal remuneration apart from out-of-pocket expenses such as cost of travel, housing accommodations and the like. It had no truly official home at the time, and in that sense was essentially a paper organization.

ALEXANDER DALLAS BACHE[2]

The Academy started with fifty charter members under the presidency of Alexander Dallas Bache (1806–1867). The number was not increased during the remaining years of the war because a generally acceptable means for electing them had not yet been developed. Bache was a great-grandson of Benjamin Franklin, an outstanding West Point trained physical scientist and engineer as well as an expert in numerous other fields. He had been born and raised in Philadelphia and always looked on that community as his home base.

An individual of quiet demeanor, Bache was a very serious student at West Point and graduated first in his class. He was greatly admired by a group of colleagues who took the time and effort to shield him from occasional sportive distractions in which they might engage. On graduation in 1825, he continued at West Point as an assistant professor for a year. This was followed by two years of engineering duty in Rhode Island, which was followed in turn by an appointment to the chair of physics and chemistry at the University of Pennsylvania. He also accepted a position at the newly created Franklin Institute, which gave him an opportunity to investigate important technical problems, such as the frequent explosion of steam boilers. In addition, he became interested in fluctuations in the earth's magnetic field and made measurements of it. He also studied the relative thermal effects of various parts of the visible and invisible portions of the optical spectrum derived from the sun.

In 1836 the trustees of the newly created Girard College for Orphans in Philadelphia offered him the position of president, which he accepted reluctantly out of a sense of duty to public service. The offer carried with it a fund that permitted him to make a two-year study of educational practices in Europe, where he was well received everywhere he traveled, partly in memory of his widely famed great-grandfather. On returning he found that the post at Girard College was not yet available because of delays in building construction. As a result he offered to make a study of educational practices in the public school system of Philadelphia, serving as head of the Central High School of Philadelphia. His offer was accepted and led to a substantial modification of the ongoing system in keeping with his European experience. In the meantime, matters continued to lag at Girard College, but the University of Pennsylvania gladly welcomed his return in 1842. He was soon extending research programs he had left behind.

A year or so later he was offered the newly vacant position of Superintendent of the United States Coastal Survey, which he accepted with his usual

Figure 1.   Alexander D. Bache

trepidation when considering a major change. Actually he found his life's mission in the new post and was very successful. He received many auxiliary appointments, some honorific, others in the line of duty. His relations with Congress were excellent although he was occasionally asked: "When will you finish the survey?" His invariable response was: "When you cease annexing territory!"

To accomplish the tasks of the Survey, Bache needed a very competent staff and was fortunate in acquiring some extraordinarily capable individuals. He found one, George Davidson, while working with the Philadelphia educational system. Davidson was the leading student in his high school class and proved to be a very helpful aide in connection with Bache's experiments. He joined the Survey on graduation and, as we shall see later, soon rose to one of the leading positions on the staff.

In his will, Alexander Bache left the sum of $40,000 in trust to the Academy with the understanding that the income would be available to his widow for the remainder of her life and could then be used for the needs of the Academy. This first significant bequest played an important role during the start-up peacetime period that followed. By 1895 the accumulation of such trust funds came to $87,000.

The Smithsonian: Joseph Henry[3]

At the time the National Academy of Sciences was formed, the most distinguished scientist in the country was Joseph Henry, the first director (Secretary) of the Smithsonian Institution in Washington, D.C.—a position to which he had been appointed in 1846. This followed a series of Congressional debates that lasted some ten years following the arrival in Washington in 1836, in the form of gold bullion, the very significant bequest by an Englishman, James Smithson, the illegitimate son of the Duke of Northumberland, that funded the start of the Smithsonian. Smithson's intent was to try to help the young democratic country become as deeply involved in scientific research as possible. When minted into American gold coinage the total bequest from Smithson totaled about $650,000—a very large fortune for those days.

Henry was made a founding member of the Academy, but had not been informed of the detailed plans of the founding committee along the way, presumably either for fear that he would dominate decisions regarding details, or might disapprove of the creation of an academy during wartime. Actually, Henry expressed concern regarding the fact that the newly formed Academy had roots in governmental action, since he feared that it might thereby become politicized. He had reason for his fears since, at the Smithsonian, he had to deal with the membership of a regency that oversaw his activities and was

Figure 2.   Joseph Henry

appointed by Congress. The Chief Justice of the United States serves *ex offi-cio* as a member of the regents and by tradition as chairman (Chancellor). The Vice President also serves *ex officio*. The membership includes three senators, three representatives from the House and six from the more general public. The latter may include heads of museums, distinguished academics, and other members at large, such as lawyers and businessmen. Service on the board of regents is regarded as a special honor, particularly among individuals close to the District of Columbia. Actually Henry was not only a skilled politician, but he quickly won the deepest respect of the Washington community and had lit-tle difficulty personally dealing with the regents. Finally he did agree to be a member of the newly formed academy in spite of the misgivings mentioned above.

Henry (1797–1878) was born in Albany, New York. While growing up he spent time both there and with relatives in the nearby smaller town of Gal-way, which has commemorated this fact by naming a very handsome ele-mentary school building after him. His advanced education was obtained at the Albany Academy, which continues to be an active institution and which had a year-long celebration of the bicentennial of Henry's birth in 1997. There he developed two great interests: science, particularly physics, and act-ing on a live stage. He was torn in making a selection between the two when he finally had to decide on a career, but swung to physics when he had an op-portunity to experiment with large electromagnets of his own making at the Albany Academy. Incidental to this, he discovered the magnetic induction of electric currents in electric conductors in 1831, a year before Faraday who published the discovery first. Both individuals, however, are fully credited with independent discovery of this phenomenon. The official name assigned to the physical unit of electromagnetic inductance is *The Henry*. Henry's ability as an actor may have stood him in good stead during the Washington years that were to come.

Henry remained at the Albany Academy until 1832, when he accepted a professorial appointment at what is now Princeton University. There he remained actively creative, particularly in connection with the development of the electromagnetic telegraph, until he was called to Washington to head the Smithsonian.

During the Civil War, Henry spent as much time as possible studying the properties of electrically operated devices that could be used for communication, often working at night in one of the towers of a building of the Institution when he was apt to be relatively free of other obligations. A private citizen noted the flashing lights one evening and rushed into the White House to inform President Lincoln that a spy was using the Institution to signal across the Potomac. The President formed a small detachment of soldiers and rushed to the tower, only to find, and meet for the first time, the great scientist busy doing what he enjoyed most. They became good friends. It is said that Lincoln once made the comment that he believed that Henry was the most intelligent individual he had ever met.

The members of the Academy investigated the problems that came to them during the war as best they could, although nothing truly outstanding in the way of innovation occurred as a result of their activity. This was a period in which most innovation centered about the activities of ingenious inventors who focused on mechanical and early electrical systems, such as steam-powered, iron-plated warships, rifled guns and improvements in the telegraph system.

When the Civil War ended in 1865, the members of the National Academy were faced with the problem of determining how it might be used in peacetime. They had models in Europe that could be employed as a guide. However, the latter had not only been established for a relatively long time but along the way had been endowed, usually by royalty and "philosophers of wealth", with significant resources such as money for running expenses and buildings where meetings could be held and records stored. In real terms, the National Academy of Sciences only possessed a charter and the interest of its founding members.

In the meantime Bache had resigned as president at the end of the war because of rapidly failing health, leaving his office in the hands of the vice president. The latter was also in bad health and in turn gave his office over to Joseph Henry. The most obvious solution was to offer the position of president to Henry who had the capacity to serve in an imaginative manner and provide something in the way of a significant foster home at the Smithsonian. He refused. It was not only that he was approaching his seventieth year, but also continued to enjoy carrying on research, particularly in connection with acoustics. Still further, he had not been greatly enamored of the surreptitious manner in which the Academy had been formed along with its links to the federal government.

JOSEPH HENRY

Finally, when all else failed, the founders unanimously elected Henry, who had been serving as acting president, to the position of president in 1868 and abandoned further search. Placed in this position, Henry must have thought deeply about the situation, granting that he had done his best to avoid it. It must have been clear to him that a long time might pass before another opportunity would arise in which the scientific community would be able to develop a viable National Academy of Sciences if the existing one were allowed to lapse. Finally, he decided to move ahead and accepted the position. He held it from 1868 until his death in 1878. The academy remained somewhat of a foster child of the Smithsonian until it entered what was truly to be its own headquarters on Constitution Avenue in 1924. Until then, most of the meetings were held at the Smithsonian.

Once in office, Henry moved with alacrity to straighten affairs. The annual election of members was to be determined by a set schedule and based primarily on the quality of the candidate's original scientific research as determined by peers who were already members. Individual members were to belong to one of a group of ten sections as determined by their primary professional interest. A pattern of officers and regular meetings was formalized. By the time Henry left office, the Academy was pointed in the direction it would try to maintain thereafter, and for the most part has. There would be major additions to the areas of science that were given official recognition as "mature", with flexible guides governing the number of members elected each year in each professional field. This flexibility was required as the country emerged as one of the leaders in scientific research in the world and the size and complexity of its scientific community increased. Also, the Academy's structure would become more complex as it added organizations it created for special purposes, such as the National Research Council (NRC) in 1916, the National Academy of Engineering (NAE) in 1965 and the Institute of Medicine (IOM) in 1970.

Many of the activities of the Academy tended to be available to an exclusive few. However the popular interest in the products of science was large in the Washington area. To bridge this gap, Henry decided to create The Washington Philosophical Society in which the results of scientific research could be discussed before a large popular audience. Membership in the Society was relatively open to any individual who had appropriate credentials and interest. The Society still exists today.

President Grant declared 1876 a year in which to celebrate the centennial of the signing of the Declaration of Independence. The Academy joined in the festivities, but did not invite any foreign academies to send formal representatives. Henry decided that the American contribution to the exact sciences was so meager at the time that it would be unwise to display our limitations. Nevertheless many members of foreign academies did attend the celebratory meetings partly to pay respect to the burgeoning nation, and partly to see how people lived in a republic. Some visitors who had been reared on Western and Midwestern lore about the country were impressed with the extent of the wild forestlands.

We do not know if the news had reached Henry by the nation's centennial year, however, that in 1875 a twenty-seven year old American, Henry Rowland (1848–1901), working very skillfully with facilities provided in the laboratories of Hermann Helmholtz in Berlin, was able to verify experimentally Maxwell's hypothesis that a varying electric field generates a magnetic field, the counterpart of magnetic induction mentioned earlier but a far more difficult matter to demonstrate. Alongside of this was the emergence of theoretical research at the world-class level by Josiah Willard Gibbs (1839–1903) in fields such as thermodynamics and statistical mechanics. The spirit needed for the cultivation of the exact sciences was alive and well in the new world and merely needed encouragement.

The transcontinental railroad began service soon after the end of the war (1869), opening the continent to relatively convenient travel. The first serious discussions regarding the possibility of cutting some form of canal through the Isthmus of Panama took place at this time. However fifty years and Herculean efforts that involved employing the best mechanical equipment and health-preserving knowledge available at the later time would be required to achieve the goal.

Events at the Academy turned out to be complex and even somewhat tempestuous at times during Henry's tenure as president. Some members who were based far from Washington felt that the content of the meetings was too meager to merit the time and expense of travel and either resigned or ignored the responsibilities of membership. Others objected to some of the individuals chosen and voiced complaints. Then there were ill-conceived attempts to form competing science academies that had to be forestalled in Congress. Henry struggled through these difficulties with well-defined purpose. Dealing with them was sufficiently taxing that the regents decided to provide the financial means that made it possible for Henry to spend a six month break in Europe where he visited laboratories, institutions and other places, particularly in England, and discussed the early stages of the development of the Smithsonian. He was given appropriate recognition as a great scientist.

There was also fear that larger sections of the Academy would come to dominate the election procedures through strength of numbers and squeeze out the smaller ones. As a result, a decision was made in 1870 to abandon the division into sections. They were, however restored in 1895. Actually, the basic difficulties were not adequately resolved until the following century when sections were assigned individual quotas that could be varied occasionally and new entities that focused on engineering and the medical professions were created.

Many members did report on their personal research at the meetings. At the time this tended to emphasize the less abstract areas of science, that is, the so-called natural sciences, particularly descriptive biology. There was little mathematics, physics or basic chemistry.

As an institution, the Academy became involved in two programs of general interest. One of the peripheral concerns dealt with an ill-fated scientific expedition to the Arctic by sea in which the ship became entrapped in ice and destroyed. The other involved an advisory role in the consolidation of several organizations that had undertaken geological and geographic surveys of the continent. As will be mentioned later, the latter led to the creation of what is now the U.S. Geological Survey within the Department of Interior (1879).

Henry's fears concerning possible dire consequences that might result from the Academy's links with the federal government have not yet been realized by it. Occasionally someone on the Hill has been sharply critical of a report issued by the Academy, or some stated matter regarding policy. On the whole, however, the government has not only been respectful of its work, but has frequently called upon it heavily for advice. This was particularly the case in the decades immediately following World War II, when our country was evolving new structures for supporting scientific research. At worst the Academy has been ignored at times when some of the members thought it might have been of substantial help in determining national policy on issues, ethical and otherwise, that had substantial scientific content.

Nevertheless, there is currently a strong feeling within the membership that the Academy should gain more independence from the government in providing support for special areas of research, such as those on which the use of government funds is restricted because of ethical issues. The membership now hopes to achieve this goal by increasing its endowment by a substantial amount. Henry would have approved, provided the fundamental principles upon which the scientific method is based, namely a careful combination of speculation and observation along with a level playing field devoid of political bias, are preserved in professional deliberations.

The Smithsonian Institution has not fared so well recently. During most of the years following Henry's death, the regents of the Institution turned to the Academy for advice in selecting a new Secretary. As a result, the Institution enjoyed the services of an unbroken chain of Secretaries who were distinguished scientists and members of the Academy. This pattern broke down when one individual who held the office, an outstanding scientist in his own field of research and a member of the Academy, produced an exhibit that caused some bitter controversy. The successor selected by the Regents, a lawyer, was not a member of the Academy. The subsequent Secretary was previously a banker. There is evidence that the breach may heal to a degree but it is too early to say what pattern will prevail in the future. In any event, and in keeping with Smithson's wishes, one would hope that the Regents will again select a well-recognized scientist with good administrative ability as Secretary.

It may be added that Henry had earned so much admiration as a result of his activities in Washington that the procession and ceremonies that developed at the time of his funeral are said to have matched those of some of the prominent generals of the Civil War.

WILLIAM BARTON ROGERS

Immediately following Henry's death, the vice president, Othniel Charles Marsh (1831–1899) a paleontologist, who possessed a professorship at Yale, assumed the office of president while elections for a new president took place. The individual finally chosen was William Barton Rogers (1804–1882), the president and founder of the Massachusetts Institute of Technology (MIT). His education was in physics and chemistry but he became an expert geologist as a result of his extensive interest in applying those disciplines to the problems of geology, including the relationship between agriculture and the constitution of soils.

Rogers was born and raised in Williamsburg, Virginia, where his father, an Irish immigrant, was professor of physics and chemistry in William and Mary College. He received his education there and was given his father's chair at the age of twenty-four on the latter's death. He became a celebrated lecturer while engaged in research. Initially, he focused on fundamental problems in physics such as the formation of dew and the behavior of currents derived from voltaic cells, but he soon became involved in geological problems, for which he gained fame.

Figure 3.   William B. Rogers

In 1835, he accepted a professorial position at the University of Virginia and at the same time was officially appointed state geologist with access to funds that would permit a state-wide geological survey. In the new post, and while maintaining a normal teaching schedule, he and his brother Henry, who was state geologist of Pennsylvania, joined forces in detailed studies of the Appalachian region that were carried out between 1835 and 1842. Their survey was both detailed and wide-ranging. For example, they marked the nature of soils, the location of hot springs and studied the solvent action of water on individual geological formations and its effects on plant growth in neighboring areas.

On the tectonic side, they devoted much attention to the rows of folded mountain chains that characterize substantial portions of the Appalachians and must have wondered about the nature of the underlying forces that brought about the folding. They studied detailed consequences of folding on individual ranges and wrote reports that drew worldwide attention. The brothers followed up their survey with many papers derived from details of the information they had gathered.

Rogers had a long-standing interest in technical education. As a southerner possessing a professorship at the University of Virginia he had well-developed plans to establish an institute of technology in the state. Unfortunately many of the students in the Southern academic institutions were in a violent mood, being much less focused on education than on the politics associated with possible withdrawal of the Southern States from the Union. Roger's wife, who was from New England, urged him to consider placing the institute in Boston where the climate for it would be much more favorable. Rogers, who was of a genial, peaceful nature, heeded her advice when a militant student "of excitable nature" shot one of Roger's close Virginia friends who opposed separation from the Union. The new institute received a state charter in Massachusetts in 1861. On the somewhat amusing side, Rogers recommended that the institute remain permanently in Boston, and avoid any suggestion of moving to Cambridge. At that time, Harvard students had the reputation of being unruly, and he was not anxious to see that characteristic develop at his technical school.

Unfortunately, Rogers, who was in his mid-seventies when he took office at the Academy, suffered from ill health during the period and it became necessary for Marsh to take over his responsibilities on several occasions. Fortunately, the unrest within the ranks of the membership with which Henry had to deal had settled down to a considerable extent so affairs within the Academy moved ahead more smoothly.

Perhaps the major outside development affecting the Academy at the time was its involvement in matters of public health. A deadly epidemic of either yellow fever or cholera broke out in Louisiana in 1878 and rapidly spread up the Mississippi River, making it clear that some form of action at the national level was needed in the event of serious epidemics. The immediate reaction of Congress was to create a National Board of Health and call upon the Academy to serve as an advisory body. Activities lapsed after the epidemic waned and the Board was eventually replaced by the Public Health Service, which evolved out of the Marine Health Service.

Rogers died suddenly in 1882 while distributing diplomas at commencement in Boston. While his scientific work was very distinguished, one of his important contributions as president of the Academy was, as mentioned above, the ability to promote peace and calm in the organization mainly through force of his genial personality.

John Wesley Powell

The period immediately after the Civil War saw extensive exploration and surveying of the lands west of the Mississippi River by a number of essentially independent groups, usually having relatively independent goals. The one that drew special public attention, in its way comparable to that received by the Lewis and Clark expedition to the West Coast (1804–1806), was John Wesley Powell's exploration of the Colorado River, carried on as a part of the study of the Rocky Mountain Area and its extensions (1871–1879). The situation called for an essential amount of unification of such exploration in order to provide coherence and continuity. The result, led to a considerable extent by Powell, was the creation of the U.S. Geological Survey, which will be discussed below.

Figure 4.   John W. Powell

Powell (1834–1902) had such an indelibly strong influence on the history of Western explorations that a brief summary of his career here seems appropriate. Both his parents were English immigrants, his father being a Methodist preacher who moved about upstate New York, where John was born, and the Midwest. His father was a committed abolitionist and gave much voice to his opinions. His son held the same views, reinforced by a visit to the South in 1860 when he was twenty-six years of age. That visit convinced him that a civil war was inevitable. In his student years, he had been intensely interested in obtaining a good education and had the capacity for it, but had to gain what he could in segments because of the itinerant life of the family. Since his father was often on the road, John frequently had to manage the family farm and sell some of its produce wherever they had settled. Fortunately there was a ten-year stay in Wheaton, Illinois, which permitted him to gain sufficient preparation to qualify for a teaching position at eighteen years of age.

Starting as a boy, Powell enjoyed roaming alone through the meadows, woods and along the streams and rivers examining and absorbing all details of the natural setting. He collected fossils and mollusks. When somewhat older he began boating, again alone, along extended stretches of the larger rivers. He joined and was an active member of the Illinois State Society of Natural History.

When the Civil War started in 1861 he enlisted as a private at once, but was soon given the rank of Second Lieutenant. He recruited an artillery company for which he was commissioned captain. He received a severe bullet wound in his right wrist in the battle of Shiloh (April 1862). The wound was poorly treated and required amputation below the elbow. It left him with periodic nerve-induced pains that were alleviated by remedial neurosurgery only late in his life.

On returning to the war, he became associated with General Grant and was commissioned first as Major of Artillery and then as Chief of Artillery. He was fully in service during the siege of Vicksburg in March of 1863, spending much of the considerable spare time extracting fossils from the protective trench-work that had been engineered by the Union Army. He was given an honorable discharge in January of 1865 when Grant was involved in the final attack on Richmond. Powell abhorred war, but gained great experience in leadership during his years of service.

Returning to Illinois, he was offered the position of clerk of DuPage County but accepted a less lucrative post as professor of geology at Illinois Wesleyan College at Bloomington and lecturer and curator of the Museum at Illinois Normal University nearby. In 1867, with aid from the Illinois State Society of Natural History, he led a trip across the plains to the Rocky Mountains with a group of sixteen "naturalists, students, and amateurs". They had a rich experience in the mountainous wilderness. General Grant helped the group obtain railroad passes and inexpensive access to government stores at army posts.

This journey was followed the next year by a longer one supported by several colleges and the Smithsonian. Powell and some of the group wintered over in the valley of the White River, a tributary of the Green River, which in turn is a tributary of the Colorado. Here Powell accomplished two important objectives. He made the basic plans for his famous journey by boat down the Colorado and began to cultivate friendships within Indian tribes, an educational exercise that eventually made him a well-recognized expert on Indian ethnology.

One of his friends, fearful of the planned journey on the Colorado River, warned him of the possible existence of impassible falls in the river. Powell replied in effect: "Rapids yes, but a muddy gritty river like the Colorado creates a relatively smooth bottom!" In fact they eventually encountered only one significant waterfall that was the product of a relatively recent basaltic intrusion.

The river journey, which was financed by a special appropriation of Congress, started in May of 1869. The group embarked in four boats onto the Green River where they had easy access and continued on to the Colorado. The rest is a part of national history. They emerged three months later on August 2nd. The press had given them up for lost and they enjoyed reading their own obituaries. Three members of the party had tired of the continuing hardships and had climbed out of the canyon somewhere midway. The trio were befriended at first by an Indian tribe, but then put to death when an Indian from another tribe mistakenly identified them for three miners who had murdered a young Indian woman. He supported his accusation by stating that it was not possible to climb out of the canyon at the place they claimed.

The detailed reports of the river journey trickled out in various forms and pieces over the next quarter century. They made it evident that Powell was not only an individual with an iron will and matching stamina in spite of his disability, but commanded the deep loyalty of those who were a match for his spirit and were willing to accept his leadership.

Powell continued these exploratory adventures for several years during which he made geological surveys of the Rocky Mountains and detailed studies of the potential uses of the western dry lands. He concluded that the latter had very limited use for agriculture because of the lack of water. Not only was it highly variable at best, but in most places, canal-fed irrigation offered the only possible solution when at all feasible. He also began to advocate federal control of water supplies originating on public lands in the West. He did his best to warn immigrants to the West to be realistic in their expectations, especially when considering farming the dry lands.

One of his special interests lay in discovering ancient rivers whose original course had been through relatively flat land and which had succeeded in retaining that course in spite of the subsequent development of mountain ranges along their path. The rivers had succeeded in cutting through the rising land as fast as it emerged.

In June of 1878 the Congress called on the Academy to carry out a study to find good ways of consolidating the many independent surveys of the West that were being made for diverse purposes. The Academy selected Powell as the obvious chairman, even though he was not yet a member. The end result was the creation of the United States Geological Survey, which was empowered to deal with all related issues. The bill was passed in 1879. The first appointment as director went to Clarence King, a noted geologist who had long-standing relations with the government and had led what became known as The Fortieth Parallel Survey. Powell as chairman had ruled himself out as a candidate because of his role in the study. Instead he was appointed to the directorship of the Bureau of Ethnology at the Smithsonian, a post he held for the rest of his life.

King detested administrative work and resigned in two years, whereupon Powell received the appointment. He tackled the position with the full force of his personality and gathered together a highly competent and loyal staff. Among his many accomplishments, he started the production of the famous Geological Survey Maps of the country, making them readily available to the population at low cost. All went well until the late 1880s when he took on the problem of trying to gain control of irrigation waters originating in public lands. Private interests that opposed his plans exercised their influence in Congress and succeeded in having his budget and staff appointments severely cut in each of two successive years. He resigned in 1894 and turned his attention to ethnology. He was one of the founders of the Cosmos Club, a gathering place of intellectuals in Washington. The early meetings took place in his home. It was probably a spin-off from the Washington Philosophical Society founded by Joseph Henry.

OTHNIEL CHARLES MARSH

After due deliberation following Rogers' death in 1882, the membership of the Academy selected Othniel Marsh (1831–1899) as his successor. He served for twelve years (1883–1895). His ancestry was composed of New Englanders of English origin. He was born in Lockport in western New York where his father had a farm near the Erie Canal. Unfortunately his mother died of cholera when he was three years of age. While growing up, he had good relationships with his stepmother and her children but was often in conflict with his father. As a result, he did not spend much time at work on the family farm, but preferred to wander about the countryside observing nature. The Erie Canal was undergoing widening in their area, and he began to collect fossils revealed by the diggings. In the course of this activity, he fell under the influence of an expert paleontologist, Colonel Pickering, from whom he learned much and began to proceed more systematically in developing his collections. Initially he took his formal education fairly casually, although he did receive sufficient certification to be able to teach in grade schools, which he did not particularly enjoy.

All this changed abruptly in 1851 when he was twenty years old. He suddenly decided that he wanted the best education possible and entered the Phillips Academy in Andover, Massachusetts. He used the first year for orientation but then took off as a brilliant scholar, winning all of the prizes and awards within his reach. On graduating he decided to further his education and entered the Classical Course at Yale University. Here he was both encouraged and supported by a very wealthy uncle, George Peabody, his mother's brother. He did creditably well at Yale, but hungered for further education directed towards natural science, his first love. In consequence, he spent the next two years at Yale in the newly created Sheffield School of Science where he was tutored by some of the best naturalists in the country.

By 1862, Marsh had determined that his goal was to obtain a professorship at one of the leading American universities, which could be aided by gaining

Figure 5.   Othniel Marsh

further experience at some of the distinguished institutions in Europe. His uncle agreed to support him in this venture, so he took off for Europe with letters of introduction from members of the Sheffield School of Science. He spent much of the next four years in Europe, principally in England where his uncle resided, and in Germany and Switzerland. In the course of this activity he came to know personally some of the great European scientists, including Sir Charles Lyell, Sir Charles Darwin and Thomas Huxley. Thereafter he spent much time in Europe, although firmly based in America. In the meantime, his uncle had decided to provide Yale with a gift of $150,000 to create a museum of natural history.

In 1866, Marsh received an appointment as professor of paleontology at the Sheffield School of Science, but he transferred to Yale College in 1879. He promptly proceeded to become the pioneering leader in American paleontology. The chance discovery of the bones of gigantic reptilian vertebrates in the western plains and in the region around the northern end of the Rocky Mountains caught his attention and he proceeded in a highly professional way, using his own funds as well as those from whatever private or public sources he could find. He hired scouts to follow up early discoveries and provided instructions so that his staff could retrieve specimens in an efficient and nondestructive manner. This involved adding to the practical science of recovery along with the science of interpretation. In the process, he displayed all the enthusiasm of an avid stamp collector, being thoroughly immersed in his medium. His findings went far beyond quantity, although he filled museums with specimens of high quality. They included, for example, the discovery of many species of dinosaurs, including relatively large flying reptiles (pterodactyls) and toothed birds. He demonstrated beyond doubt that the horse had emerged first in the Americas and had undergone its early evolution there. It then migrated to the Eurasian continents over some land bridge before becoming extinct in its original homeland.

One of his important studies involved determining the ratio of brain size to bone weight for the dinosaurs and their ancient reptilian relatives. He had noted that the brain cavities of these creatures were relatively tiny compared to those of modern mammals of comparable size. A systematic study of the fossil brain cavities of ancient mammals demonstrated, however, that their brain/bone ratio also was small in the early stages of evolution, but increased substantially in the course of geological time, indicating a gradual increase in intelligence. One might conclude that the main occupation of both the dinosaurs and the early mammals was the search for food and procreation.

Marsh had begun his career as a creationist, as was common at the time, but noted that important European colleagues took Darwin's treatise of 1859, which presented the theory of evolution, very seriously. His own work soon convinced him of its general validity. In 1880 he was much pleased to receive the following letter from Darwin (1809–1882):

Down, Kent, August 31, 1880.

My dear Prof. Marsh,—

I received some time ago your very kind note of July 28th, and yesterday the magnificent volume [Odontornithes]. I have looked with renewed admiration at the plates, and will soon read the text. Your work on these old birds and on the many fossil animals of N. America has afforded the best support to the theory of evolution which has appeared within the last 20 years . . .

With cordial thanks, believe me

Yours very sincerely
(Signed) Charles Darwin

Marsh was made Vertebrate Paleontologist of the Geological Survey in 1882 and held the position until his death. He was of great help to Powell during the latter's happier years as head of the Survey. When he became president of the Academy in 1883 he continued the search for skeletons of ancient vertebrates that were being found in the western lands of the continent.

Marsh developed very definite views about the relationship between the government and the Academy. These were expressed in the following terms in 1889:

The question has arisen, shall the Academy, in addition to the duty of giving advice when asked, volunteer its advice to the Government? Members of the

Academy have urged this course at various times in the past, and during the present session the question has come up again for decision. My own opinion on this subject, after careful consideration, is against such action. The Academy stands in a confidential relation to the Government as its scientific adviser, and in my judgment it would lose both influence and dignity by offering advice unasked.

In appointing committees on the part of the Academy, I informed them that the proper province of the National Academy is not merely to make a technical examination of any case, but especially to bring out the scientific principles involved in the investigation, as basis for future use.

Marsh, a bachelor, did not always possess a warm friendly personality and frequently carried on feuds with his scientific rivals, some with closely controlled ferocity. He was, however, also known to be very convivial at gatherings with close friends. He was offered another six-year term as president in 1895 since he had been an excellent leader in the managerial sense, taking the post quite seriously. He refused, probably because he sensed some underlying unrest within the membership on a number of issues, such as the absence of sections.

Maxwell's Equations

It is worth mentioning at this point that Marsh's period in office coincided with the emergence of one of the great technological revolutions that is based directly on fundamental research, essentially at the academic level, namely disclosure of the inter-relationships between the force-fields involved in electro-magnetic phenomena. While the simple electric form of telegraphy that depended on the magnetic fields generated by electric currents had been exploited since the 1830s, James Clerk Maxwell's development in the 1860s of the equations governing the detailed interrelations between electric and magnetic fields permitted much broader utilization of electromagnetism as a result of more precise understanding. The construction of motors and generators could now proceed on the basis of sound principles.

Maxwell had pointed out that visible light is probably electromagnetic in nature. By the late 1880s Heinrich Hertz, prodded by Helmholtz, was producing electromagnetic waves of arbitrary length in the laboratory. Whereas North America had heretofore been mainly a follower of Europe in the utilization of science-based technology, it was now prepared to compete at the frontier in this new area as individuals such as Thomas A. Edison and the Canadian Reginald A. Fessenden emerged among leaders on a global scale. Edison gave a demonstration of his phonograph at the annual meeting of the Academy in 1878 at the age of 31. He was eventually elected a member in

1927. He established the first industrially oriented research laboratory in the country in New Jersey in 1876.

It should be added that Marsh's term in office also coincided with the period in which professionally trained chemists began to take over analysis, testing and control of many manufacturing fields, such as those dealing with ceramics, metallurgy and the preparation of pigments, paints, enamels and dyes. Hitherto, these had been in the hands of what might be termed 'traditionalists' who had based their products on generations of testing by trial and error. The scientific approach was extending everywhere, not least to agriculture.

Henry Rowland

It was, however, during this period that Henry Rowland, on retiring as the first president of the American Physical Society (1899), lamented the fact that so much effort within the American scientific community was devoted to practical applications rather than to the discovery of fundamentals. In the lecture, he provided a brilliant summary of the leading unsolved problems of the time. He hoped for a day when there would be more balance. That day was not far off, although Rowland was not to see it as a result of an early death caused by then untreatable diabetes. Anticipating his death, he had diverted his interests to the development of the Teletype machine in order to provide an income for his family.

Figure 6.   Henry Rowland and his engraver for producing optical gratings.

The growth of general interest in science in the country is indicated by the fact that Alexander Graham Bell and a colleague were able to start and sustain the journal *Science* in 1883. It was eventually taken over by the American Association for the Advancement of Science, which had been founded in 1847.

The Congress soon realized that in one way or another it was not only being compelled to make decisions that involved scientific input, but also that the federal agencies were beginning to add many scientifically trained individuals to their staffs. The question arose: "Should there be consolidation?" The matter was referred to the Academy in 1884 with the recommendation that it organize a study committee with very broad membership, not limited to that of the Academy. With the help of Congress, Marsh selected such a committee whose studies were carried on between 1885 and 1887. The official chairman was Senator William B. Allison, of Iowa, after whom the committee was named. The principal conclusion drawn was in opposition to consolidation. It was felt that the need for scientific advice would be increasingly diverse and could not be encompassed within a single organization. Flexibility was important. Representative Theodore Lyman (1833–1897) of Massachusetts, a member of the Academy, provided much guidance.

Lyman was a well-known naturalist, having published a number of papers describing the details of native plants and other matters. The popularity that won him a seat in the House stemmed from his military service in the Civil War. He was on the staff of both General Meade, the victor at Gettysburg, and General Grant. He was involved in a number of battles, including Vicksburg and the final campaign leading to Appomattox. He became a lifelong friend of Meade.

Perhaps the second most important conclusion that emerged from the study centered about the need to standardize commercially used weights and measures. Previously this responsibility had been linked to the Coast Survey. The new review ultimately led to the creation of the National Bureau of Standards (1901). At start the Bureau was attached to the Treasury Department but now resides in Commerce. Attempts to have the metric system adopted universally in the country, for use beyond scientific measurements, failed then and have continued to fail since.

OLIVER WOLCOTT GIBBS

The next choice for president was the chemist and charter member of the Academy Oliver Wolcott Gibbs (1822–1908), who had refused the office earlier, but was now willing to serve between 1895 and 1900. He had an excellent foundation in analytical and inorganic chemistry obtained at home and

Figure 7. Wolcott Gibbs

abroad and was deeply committed to a research career. His first opportunity after returning from studies in Europe came through a teaching position at City College in New York where he obtained access to laboratory facilities for both students and research. He remained there productively for seven years. In 1863, he was offered the Rumford Professorship at Harvard, a prestigious post endowed in his will by the notorious Tory scientist and adventurer Sir Benjamin Thompson as somewhat in the nature of a peace offering to his native land for his role in opposition to independence in the Revolutionary War.

At Harvard Gibbs had good access to research facilities and research students initially, but these were denied him after 1871 as a result of the reorganization of the university and its adjuncts. It is strongly suspected that the administrations of both Harvard and Yale desired to downgrade the status of science as a university discipline at the time and give stronger emphasis to the classics. Gibbs' research activities were ultimately confined to those in a private laboratory he had developed at his summer home. He gained special fame for his work on the platinum family of elements.

John von Neumann once whimsically commented that science seemed permanently secure at Princeton University. Each time an incoming administration stated that it planned to downgrade science in favor of some other field of endeavor science had managed to become stronger.

While Gibbs' service to the Academy was somewhat relaxed during his five years in office, he did take a strong interest in matters of conservation of natural resources. He was appalled at the wanton exploitation of the nation's forest and mineral lands that was occurring as a result of the activities of the lumber and mining companies, and he worked closely with Congress to produce correctives. For example, he was effective, with the help of advisory committees composed of distinguished individuals, in supporting legislation that created a system of National Forest Preserves in 1901. Initially the act focused on Western forests, but was extended to the Appalachians in 1916.

Muir-Pinchot Debate

This was the period in which a great debate arose concerning the actual manner in which the public forest preserves should be maintained. John Muir supported the policy of "forever wild" whereas Governor Gifford Pinchot of Pennsylvania believed that generally and in the long run we should follow the European practice of carefully grooming public forestlands in order to reduce the risk from uncontrolled fires and plant disease and to optimize the quality of lumber—a valuable resource. This issue has not yet been resolved, being still a matter of open controversy. Muir, supported by President Theodore Roosevelt, succeeded in helping to create the Western national parks. He appealed broadly to the public through a series of brilliant essays on the subject, and overcame opposition from the lumbering interests.

ALEXANDER AGASSIZ

The next choice for president, made in 1901, was Alexander Agassiz (1835–1910), the scientist son of the Swiss immigrant Louis Agassiz (1807–1873). The latter was a charter member of the academy who had enjoyed a brilliant career in United States after gaining fame for research in Europe on the study and classification of fossil fish. He had also provided the first reasonable, scientifically comprehensive understanding of the great ice age and its distinctive effects upon geological features produced by the motion of the great glaciers of the past. The latter revolutionary work led, for example, to the explanation of the U-shaped nature of Yosemite Valley, first given by a young John Muir working as a temporary lumberman in the valley. Muir had heard Louis Agassiz lecture on the physical effects of the movement of glaciers while a student at the University of Wisconsin but had great difficulty selling the concept to the well-established American geologists of the time.

Louis came to United States to deliver lectures in Boston in 1846 and, liking the challenging environment, decided to stay after receiving an appointment as a member of the Harvard faculty. He was lionized in America as a world-class European scientist. He was soon omnipresent on the American scientific scene and was universally regarded as one of the particularly "elite"

Figure 8.   Alexander Agassiz

members of the scientific community, a status to which he did his best to conform. He became a regent of the Smithsonian and received many other honors. The elder Agassiz exhibited one major idiosyncrasy: he had significant doubts about Darwin's theory of evolution. His son did not share these doubts. The leading naturalist of the time, Alexander von Humboldt, regretted Louis Agassiz's shift from fossil fish to glaciers since he had more interest in fish. Apparently Agassiz had very acute vision and could distinguish minor differences between fossilized species.

Alexander Agassiz, who served as president of the Academy for six years, had for the most part been raised and educated in Europe. However, on settling in United States, he accepted a position at the Museum of Comparative Zoology at Harvard, which he retained throughout his career, ultimately becoming its director. As a significant side interest, he became involved commercially for a period in the development of the Calumet and Hecla copper mines in Michigan. He found them poorly managed and on the verge of being bankrupt. He entered the picture as a diligent dedicated businessman who merged the two mines and introduced much more efficient equipment. He became a young independently wealthy scientist who was free to pursue any interest or activity he desired. He was devastated by the early death of his wife and decided to change his style of life. He took full advantage of the freedom going with bachelorhood and roamed the world systematically on a variety of scientific missions of his own choosing. He accepted the appointment as president of the Academy with the understanding that he would retain much of his personal freedom of action.

Figure 9.   Louis Agassiz

President Grover Cleveland, who previously had little interest in science, became enamored with the subject after making the acquaintance of Alexander Agassiz and offered him the position of Scientific Advisor to the President of the United States in his administration. Agassiz begged off on the basis of age and health, but his real reason probably lay in the fact that the demands of the post would have disrupted well-laid plans for his personal research.

There was some unrest and associated disputation concerning the relative magnitude of the roles that the government-based scientists in Washington should play in Academy affairs. This issue seems to have vanished by the 1920s as the national community of academic scientists grew very large and self-assertive.

Revolution In Physics

Early in the 1890s, following the development of electromagnetic theory, many individuals thought that the field of physics had finally attained maturity, and that future activity in it would merely involve filling out what might be called connecting gaps. Actually, the world of physics was effectively reborn during the relatively short period between 1895 and 1907 in which Gibbs and Agassiz held office. The field started on new journeys that have continued to provide expectations of major unforeseen revelations up to the present. The members of the Academy approached 1913, its semi-centennial year, with the realization that the scientific endeavor was still very young and

Figure 10. Albert Einstein circa 1910

that great discoveries lay ahead, some important for enlightenment, others for the advance of technology.

Agassiz's term of service is distinguished by the addition of a section of the Academy in the field of psychology.

IRA REMSEN

The Academy selected New York born Ira Remsen (1846–1927), an organic chemist, to lead the institution through its semi-centennial anniversary in 1913. At the time, he was serving as president of the Johns Hopkins University, which he had joined as its first professor of chemistry in 1876. As was customary at the period, part of his education was carried out in the United States, but he continued it by spending several years in Europe where he learned much while working there in some of the relatively advanced research laboratories.

On returning to United States in 1872, he chose a faculty position at Williams College in Massachusetts on the prospect that it might soon start a graduate school and allow him to have graduate students and a suitable chemistry laboratory. When this did not materialize, he decided to accept a position at Johns Hopkins University, which was planning to pattern itself after the German universities of the period. He soon built up a fine reputation as a research chemist, teacher, and administrator. His laboratory was a prolific

Figure 11.　Ira Remsen

source of well-prepared organic chemists. As a result, he was selected as the second president in 1901 on the retirement of the founding president Daniel Coit Gilman. In the course of his academic career, which also involved a period as president of the University of California, Gilman had exerted a revolutionary influence on higher education in the country. He served as the first president of the Carnegie Institution in Washington, D.C., founded in 1902.

Included in Remsen's extensive chemical research was the discovery of the super-sweetener saccharine, made with a colleague.

Control of Commercial Trusts

This was a period in which much public concern arose regarding the growing power of the great industrial trusts. There was fear that they might gain too much political and economic control in the nation. The discussion ended in the formulation of antitrust laws that have acted to prevent the development of monopolies, or to regulate their actions when monopolies are inevitable, as was essentially once the case for the telephone system. This was also a period in which the so-called muckraking reporters began to raise questions about the conditions under which most of our everyday foods are produced. Again, new regulations were introduced, most of which have been expanded and tightened in the intervening years. The first Pure Food and Drug Act was passed in 1906.

In the same period, the Congress noted once again the ever-increasing role that science and scientists were playing in the various governmental agencies and requested the Academy once again to determine if a significant amount of consolidation was now reasonable. To lead the task in this instance, Remsen selected Dr. R. J. Woodward, the incumbent president of the Carnegie Institution. Again the committee decided that real duplication in scientific effort was very limited since different agencies had different problems and goals. It did, however, go so far as to recommend that an appropriate watchdog committee be created. Through mischance the matter never reached the legislative stage and no significant action followed.

Semicentennial Celebration

In preparation for the actual celebration of the Academy's fiftieth birthday, Remsen commissioned a semi-centennial history to be prepared by Frederick W. True[4], Deputy Secretary of the Smithsonian. The volume generated by True did appropriate justice to the history of the Academy since its start in 1863.

The celebration itself was devoted to both feasting and lectures. President Remsen expressed the hope that the government would turn to the Academy more frequently for advice on some of the many problems that had scientific content and held a bearing on national well being. He also hoped that in the future the presidents of the Academy would have the advantage of living in or near Washington in order to stay very close to the affairs of the Academy.

George Ellery Hale (1868–1938), the astronomer (who had been elected to the Academy in 1902 and was deeply immersed in its affairs), emphasized that the period had arrived in which the Academy should have a home of its

Figure 12. George Ellery Hale

own and that he hoped the time was close at hand when it could be expected to raise the funds to obtain one.

In the general framework of science, the semi-centennial year was distinguished by the emergence of Niels Bohr's quantized theory of the hydrogen atom. Ernest Rutherford had demonstrated in 1911 that the atom contained a relatively small nucleus in which most of the mass and all of the positive charge was concentrated and that the companion electrons were for the most part outside the nucleus, occupying a volume of atomic dimensions much larger than that of the nucleus. Bohr demonstrated that these facts could be rationalized to a degree by abandoning major concepts of classical physics and introducing discrete energy states for the electrons. It had become clear that Planck's constant dominated the atomic world.

Prediction of Neutron; Harkins-Chadwick

The chemist Professor William D. Harkins (1873–1951) of the University of Chicago wrote several papers during this period pointing out that the nucleus probably contained a neutral particle with a mass close to that of the proton. On returning to the Cavendish Laboratory immediately after World War I, the English physicist James Chadwick (1891–1974) made a test of this hypothesis by bombarding a number of light elements with the most energetic alpha particles available, in the hope of dislodging such a particle if indeed it existed. For better or worse, there was no beryllium conveniently available to him at the time or the study of neutron physics would

Figure 13.   William D. Harkins

have begun a decade earlier than it did with untold consequences, scientific, technical and political.

Chadwick's opportunity finally came in 1932 when the French physicists in Paris reported the production of a strange neutral radiation as they began using their newly constructed cyclotron. Chadwick immediately identified it with the particle he had sought earlier in the previous decade. He proceeded to produce it with the use of a radium-beryllium source.

In a public address, Rutherford acknowledged Harkin's role in helping to stimulate the search for the neutron.

President Remsen resigned both as president of the Academy and as president of Johns Hopkins soon after the ending of the celebration.

WILLIAM HENRY WELCH

The next president, elected in 1913, was William Henry Welch, M.D., (1850–1934), undoubtedly the most distinguished member of his profession in the country at the time. He came of a long-standing medical family but his early interest as a student was in the classics. On graduating from Yale in 1870, he decided to apprentice in medicine with his father and soon concluded that American medicine was very backward in its use of science. As a result, he decided to formalize his education by obtaining an M.D. degree from the College of Physicians and Surgeons in New York (1875) and interning at Bellevue. He then spent two years in Europe gaining direct experience from the advanced clinics there, specializing in pathology (1876–78).

Figure 14. William Henry Welch

Figure 15. Simon Flexner

On returning he began to teach at Bellevue during the period between 1878 and 1884, but hoped for a position at the Johns Hopkins University. This finally came in 1884 when Dr. John S. Billings, who was organizing the university hospital, asked him to become professor of pathology. He now found himself in very stimulating company and continued to rise ever higher in his profession. He was one of the key individuals who helped create the Johns Hopkins Medical School and Hospital, which began operation in 1889. It soon gained an outstanding reputation as the model for others to follow because of its emphasis on the scientific aspects of medicine, as well as for the quality of its clinical work.

As a bachelor, Welch had much freedom to travel and greatly enjoyed participating broadly in the international scene. He was trapped in Europe for a full month in 1914 at the outbreak of World War I.

One of Welch's many meritorious side tasks was to head the advisory committee for the development and staffing of The Rockefeller Institute for Medical Research when it was created in New York City at the turn of the century through the benefactions of John D. Rockefeller, Senior. In fact, Welch exerted strong guidance at both institutions, selecting one of his young former students, Dr. Simon Flexner (1863–1946), noted for his involvement in fundamental medical research, as the first director of the new Rockefeller Institute.

Simon Flexner

Flexner was born in Louisville, Kentucky, and received an M.D. at the University there in 1889. He was much interested in the scientific side of medicine and was to an extent self-taught at this early stage. He next went to the newly opened medical school of the Johns Hopkins University to study pathology where he was soon recognized as an unusually gifted investigator and was invited to join the staff. In this new environment, he began to follow serious epidemics and sought preventative vaccines. For example he searched for a vaccine for cerebrospinal meningitis and eventually succeeded. He identified a strain of bacillus that plagued the city of Manila in the Philippines. He struggled very hard to find a vaccine for poliomyelitis, which continuously emerged in portions of the United States. Unfortunately he selected a very obdurate form of the virus in his experimental work, or we might have had an effective vaccine for polio in the 1920s. He did, however, lay the groundwork for such research, which was finally successful in the 1950s. Flexner agreed to take the post at the new institute under the condition that the founders provide funds for a research hospital. His personal research there continued on the quest for vaccines for significant diseases.

He was joined at the Institute in the study of epidemics by the very ingenious Japanese scientist Dr. Hideyo Noguchi (1876–1928). The latter died of yellow fever in West Africa while studying the origin of that disease. One of his most notable discoveries was the existence of live spirochetes in the brain of a deceased individual who had suffered an advanced form of syphilis. In 1976 Dr. Frederick Seitz and his wife were invited to a ten day gala in Japan devoted to a celebration of the centennial of Noguchi's birth. He is regarded there as the first distinguished Japanese scientist to become involved in Western medicine in a truly creative way. Members of the Japanese royal family served as friendly hosts for part of the extended occasion.

Figure 16.   Hideyo Noguchi

This was a very active period of technological development for United States with the result that the Academy was called upon to offer help on some major problems, including some that emerged during the creation of the Panama Canal, a task that had baffled the French engineers in 1889 for a variety of good reasons. On another frontier, a question arose as to whether the seal herds that gathered in Alaska, which the United States had purchased from Russia in 1867, were being properly managed. The Academy was called in as advisor and helped stabilize the situation by introducing regulatory controls of the harvesting.

World War I: The National Research Council

When World War I broke out in August of 1914, President Wilson became personally concerned that some as yet unforeseen event might cause the United States to enter it. As a result, he decided in 1915 that our country should take steps toward "preparedness" to begin to cover the possibility of involvement. He discussed the matter somewhat privately with a number of prominent citizens and professional groups and on the whole received enthusiastic backing.

George Ellery Hale, who was foreign secretary of the Academy at the time, responded to President Wilson's charge with much enthusiasm and, with the support of colleagues, formed a committee in 1916 to review the topic in depth. Since he realized that many of the problems that might arise in wartime would be of great interest to the professional engineering community, Hale formed a close link with the Engineering Foundation, a fairly centralized organization that served some of the interests of the engineering community as well as other engineering-related organizations. Using private money at first, provided in significant part by the Engineering Foundation, the Academy began to analyze its advisory structure and policies. It quickly realized that it could not be effective on the scale that would be necessary in a now highly industrialized world if it limited membership on specialized advisory committees or teams to members of the Academy, although many of the latter who possessed much expert knowledge and qualities of leadership could be indispensable. It would need to involve many individuals well outside the Academy's membership if it were to provide broadly useful advice.

In brief, the Academy needed an auxiliary organization that drew advisors of many stripes from many sources. As a result, it created in 1916 the National Research Council, an advisory structure whose areas of interest not only included those of the principal professions of the Academy members, but also encompassed fields much closer to ongoing technology that did not have representative experts within the membership. It follows that the number of non-members who would be active as advisors within the National Research Council could be expected to outweigh the number who were members, as proved to be the case. Hale accepted the chairmanship of the new organization.

The new addition to the Academy's structure was greeted with enthusiasm by scientists and engineers as well as by segments of the medical profession. It rapidly became a very active organization, first before our entry into World War I in April 1917, but especially after entrance. In the meantime Hale, who had labored very strenuously to achieve success, had kept President Wilson informed of the actions that were taking place. The latter was very pleased and offered a few suggestions concerning the plans. The President eventually (March 1918) issued an executive order making the National Research Council an official component of the Academy, with permanent status.

CHARLES DOOLITTLE WALCOTT

When the United States entered the war, the Academy's president, William H. Welch, decided to accept a high-ranking position in the Army Medical Corps and resigned from his Academy post. In the quest for a successor to Welch, who retained close relationships with the advisory structure of the Academy, the search committee called upon the Vice President, Dr. Charles D. Walcott (1850–1927), who was ongoing Secretary of the Smithsonian. In addition to being one of the world's great paleo-geologists and a very capable administrator, Walcott possessed knowledge in depth and breadth regarding who and what were important for achieving significant goals in the Washington community. He probably can be regarded as the spiritual successor to Joseph Henry. His actions in relation to the evolution of the Smithsonian were undertaken quietly, without fanfare, but he left the institution richly endowed with new buildings, museums and special collections. His remarkable intuition is indicated by the fact that he established a committee on aeronautics as early as 1897, six years before the flights of the Wright brothers, as a result of witnessing the experiments on flight of Samuel P. Langley that were being supported by the Navy.

Walcott had been raised in upstate New York under circumstances in which he was able to have only ten years of formal education. He decided, however, that he wanted to be a scientist. A chance meeting with an older individual who was collecting fossils attracted his interest and he began a collection of his own, mainly of trilobites, while absorbing the professional literature. By age 16 he was highly experienced and decided to focus on Cambrian fossils. He made an agreement with a farmer to work part time for room and board

Figure 17.   Charles D. Walcott

and succeeded in gathering a fine collection of invertebrate fossils which were sold to Harvard University. His remarkable abilities and professional approach attracted the attention of Louis Agassiz, who made preliminary plans to have him enter Harvard. Unfortunately, Agassiz died before the plan could be put into effect, so in 1876 he accepted a position in the New York Geological Survey, then under Professor James Hall who had gained much admiration for Walcott.

As mentioned earlier, the U. S. Geological Survey was created in 1879. Hall successfully recommended Walcott for a position there. The latter extended his studies over large parts of the country in the new post and was made head of the division of invertebrates in 1882. On one occasion he ventured deeply into the Grand Canyon quite alone and often followed dangerous trails. He was becoming a distinguished master of his profession.

When Powell resigned his position as head of the Geological Survey after what had become a stormy period in Congressional relations, Walcott was selected to be his successor. This occurred with the full agreement of his colleagues since his special diplomatic skills in handling negotiations in the Washington environment were now well recognized and appreciated. He held the position until 1907 when he was appointed Secretary of the Smithsonian.

One of Walcott's great discoveries occurred in 1907 while on what was for him a refreshing research investigation in the Canadian Rocky Mountains. While traveling through the Burgess Pass he noted an out of place specimen of Cambrian shale on the trail. A study of the item demonstrated that it contained a remarkable fossil. A diligent search on the spot led to the discovery of the source, which turned out to be very rich in fossils of fauna from a special transition period in evolutionary history. Walcott sighed and said that the research there was for fortunate younger individuals.

Walcott demurred at first in his discussions with the search committee for a new president of the Academy because of his age (67), but finally agreed. He remained in office for six years, until 1923. As chairman of the National Research Council, Hale could not have had a more effective colleague heading the Academy during his most strenuous days. The two worked as a very well coordinated team both during and after the war, Walcott deferring to Hale when matters of publicity were concerned. Hale was prone to overwork to the point of exhaustion and suffered periodic nervous breakdowns as a result. For example, he apparently suffered such breakdowns in 1910, 1913 and 1921. The last was sufficiently severe that he gave up all pressure-producing administrative posts and devoted his subsequent career to scientific research on Mount Wilson and the design of a 200–inch telescope that was eventually established on Mount Palomar. Except for the problem of health, he probably would have followed Walcott as president of the Academy.

By the end of the war the various activities carried on by the National Research Council were viewed for the most part as having been highly successful. It had brought into service in a productive way both individuals and groups that otherwise might not have been able to contribute to the war effort in as effective a manner as they did. For example, it provided a vehicle that permitted imaginative individuals to make important contributions to fields such as sound ranging and sound amplification for the purpose of locating enemy artillery or early detection of approaching aircraft. Or to develop special systems of optical filters that would help reveal camouflaged systems. At the opposite level, the organization made it possible for technical and managerial staff from different industrial organizations that normally might be in competition to work together as partners instead in advancing essential new areas of technology, such as aeronautics, sonar, and radio communication.

Some brilliant private inventors, such as Reginald Fessenden, complained that such mixed industrial teams tended to abuse the rights of individual inventors. Although he had led the way in the invention and use of sonar, he was never asked to participate in the extensive wartime activities of the Council in the field for unknown reasons.

Foundation Awards

In any event, two of the large private foundations felt that the Academy should be rewarded in a significant way for its wartime enterprises. The first to act (1919) was the Carnegie Corporation, which provided a gift of five million dollars, a portion of which was to be used for the construction of an appropriate building in the District of Columbia to serve as headquarters for the Academy. The remainder was to be used as working endowment. Actually, early plans for some such gift, probably more modest, had been initiated in conversations Hale had with the Corporation prior to the war, but the outstanding success of the work of the National Research Council undoubtedly accelerated the actions of the Carnegie Corporation and enhanced the magnitude of the gift that was eventually made.

Meanwhile Robert A. Millikan, (1868–1953), then in the transition from the University of Chicago to the California Institute of Technology and deeply interested both in Academy affairs and the general promotion of science on an international scale, helped prepare legislation that permitted the Academy to own land in the District. With this start, he raised sufficient funds to acquire a parcel, actually a city block, near the Lincoln Memorial on what was then called B Street but renamed Constitution Avenue in 1931. Millikan did much in the 1920s to help promote the creation of the International Scientific Unions, which came into being in the early 1930s.

With the help of the Commission of Fine Arts of the District of Columbia and his own broad experience, Hale selected Bertram G. Goodhue (1869–1924), a deservedly celebrated architect, to design what became one of the most distinguished buildings in Washington. The basic plans for the structure were developed by April of 1921, and the building was completed in 1924, just in time for the planned dedication on April 28th. A great deal of thought went into every detail of the structure, well beyond the organization of space and the use of the finest construction materials available. The exterior presents a magnificent marble façade to Constitution Avenue with an apt quota-

Figure 18.   Robert A. Millikan

tion from Aristotle. The halls, meeting rooms and the numerous embossed bronze doors are decorated with items related to the history of the Academy, of the intellectual world abroad or symbolic of the history of science since the time of the Greeks. On display was scientific equipment including an operating Foucault Pendulum.

President Coolidge and many other notables both within and without the membership of the Academy were present at the dedication. Albert A. Michelson, the newly elected president of the Academy, as well as Walcott, participated actively in the formal ceremonies. To accommodate the complexities of the meeting as well as the size of the assembled group, the large central hall was modified to a degree by the addition of various wooden structures of high cabinet-level quality. These remained in place until the centennial celebration of the Academy in 1963 when the hall was completely refurbished. The "temporary" additions had been used occasionally in the interim.

A young architect, Wallace K. Harrison (1895–1981), who was starting on his career, had the good fortune to become somewhat of a senior apprentice on Goodhue's staff and played a major role in the design of the building. He was sufficiently stimulated by the experience that he wrote an extensive article on the construction of the Academy building in the professional journal *Architecture* in 1924. He was to play a major role in the extension of the structure in the next fifty years.

As the Carnegie Corporation was providing a home for the Academy, the board of the Rockefeller Foundation realized that the United States was now well on the road to becoming an advanced scientific nation. In order to

Figure 19.   Albert Einstein circa 1921

be certain that the most capable young scientists who had recently obtained a doctorate or its equivalent had a good opportunity to extend their period of internship further, the Board decided to provide the Academy with funds sufficient to permit a reasonable fraction to have an additional period of concentrated study and research at an institution of their own choosing. The National Research Council administered the fellowships.

Most such fellows selected major centers in Western Europe. The National Research Fellowships were not only very popular, but rapidly helped close the creative gap in science that had existed on the two sides of the Atlantic Ocean. What might have been left incomplete in the way of closure was bridged in the 1930s by the flight of refugees from Europe to the United States ordained by Hitler and Mussolini.

The fellowship grants were continued over a thirty-year period, until federal funds became relatively free. At first those in the physical sciences were favored to a degree as a result of the quantum revolution, but Warren Weaver (1894–1978), a wise and perceptive counselor newly arrived in the Rockefeller Foundation in 1931, foresaw the revolution that was to come in the biological sciences, and convinced the grantors that they should provide more balance.

In the meantime, the Academy, in an effort to express its gratitude to the engineering community, created a section on engineering. Nine existing members of the Academy who had strong interest in the applications of science joined the new section to form a base. One was Frank Barton Jewett, president of the Bell Telephone Laboratories.

One of the major transformations resulting from the participation of United States in World War I was its rise into the family of major industrial nations in a broadly competitive way. Not least, it was now a significant contender in the field of industrial chemistry.

Versailles Treaty

The victorious European allies were aghast at the massive destruction that had finally been achieved as a result of the continuation of centuries of internecine rivalries and warfare. In meetings held in Versailles, they chose to inflict very vindictive peace terms on their former enemy countries with heavy indemnities. The United States rejected that peace plan in spite of the pleas of President Wilson to accept it and made separate peace agreements once Germany and Austria had adopted democratic governments. The Versailles Treaty included the stipulation that the scientists in the enemy countries would not be allowed to attend international scientific meetings in the future.

Fortunately this provision did not prevent Albert Einstein from visiting the United States in 1921 on a philanthropic mission. In addition to being received warmly, he gave several scientific lectures in the East and Midwest. It did, however, completely disrupt the prewar pattern of international scientific congresses and eventually led to protests by the countries that had been neutral in the war and by the United States. Niels Bohr, the esteemed Danish scientist, made the statement: "If Einstein cannot go to the meetings, I cannot go!" A crisis was at hand. It was finally resolved in 1931 by the creation of international scientific unions in each of the major fields of science and under the overall governance of The International Council of Scientific Unions (ICSU). In the meantime, the clause in the Versailles Treaty that had restricted international scientific travel was rescinded.

The German economy collapsed in the mid-1920s as a consequence of the monetary burdens placed on it, causing great stress within the central European community as a whole. This generated considerable concern in English and American economic circles for fear of an eventual effect on international trade. In an attempt to help remedy the situation the two countries made a deliberate effort to bolster the German government's struggles through loans. It is debatable how much influence the collapse of the central European economy had on the generation of the world-wide depression that started in 1929, but it did, unfortunately, help the rise of Hitler and other dictators.

ALBERT ABRAHAM MICHELSON

As mentioned, Charles Walcott stepped down just as the Academy was beginning to occupy its new quarters and Albert Michelson (1852–1931) took his place. Michelson had been born in what is now Poland, but his parents had decided to come to the United States when he was still a child. They settled in California and Nevada where his father had considerable success while serving as one of the purveyors to the mines in the area.

Young Albert had his basic education in the schools of San Francisco. On reaching college age in 1869 he decided that he would like to go to the Naval Academy as an appointee from the state of Nevada where the family was based at the time. The story goes that three competitors for the position became involved in a tie and it was not clear what would happen next. Apparently young Michelson went to Washington on one of the newly operating transcontinental trains and made an appointment to see President Grant. The latter listened to his tale and liked what he heard. As a result, the President settled the issue in favor of his visitor. Michelson did well at the Naval Academy

Figure 20.   Albert A. Michelson

and became a teacher there. Along the way, he managed to spend a period working in Helmholtz's laboratory in Berlin where he became deeply involved in the frontiers of European science and honed his talents.

Soon thereafter he left the navy and took a position (1881) at what was then the Case School of Applied Science in Cleveland. There he began to carry out the experiments upon which much of his great fame rests. Using ingeniously designed optical equipment, he attempted to measure the velocity of light relative to the presumably static ether as the earth rounded its course. The most revealing experiments of the time were carried out in 1886 with Professor Edward W. Morley (1838–1923), a colleague. The more carefully they made the measurements, however, the less evidence they found for anything resembling a stationary ether. The velocity of light as measured in the laboratory appeared to remain constant for all directions of the earth's motion. This result caused much controversy, but was finally resolved when Einstein proposed the special theory of relativity as a result of his very acute insight, linked with the fact that Maxwell's equations are invariant under the Lorentz Transformation. Initially, the transformation was regarded to be no more than a mathematical device that H. A. Lorentz had introduced as a useful helpmate in solving some problems in electromagnetic theory.

The best evidence seems to suggest that Einstein was not actually aware of the Michelson-Morley experiments at the time he proposed the special theory of relativity. Apparently, even in his precocious teenage years, he had concluded intuitively that the speed of light was probably invariant in any uniformly moving frame of reference.

Edward Morley

It should be added that Morley's previous career had not been in physics but in chemistry, with astronomy as a principal side interest. He was well known for precision measurements of atomic weights and of very small variations in the oxygen/nitrogen ratio of the atmosphere under varying conditions of weather. His initial involvement with Michelson came about through the fact that his laboratory was found to be the best place on campus to carry out the experiment. Moreover, Michelson was pleased to have the cooperation of such a good scientist. In 1898 Morley repeated the experiment with Dayton C. Miller, a colleague. The results were the same within experimental error. Morley was a central figure of attention when he attended the Congress of Physics in Paris in 1900 where Lord Kelvin questioned him closely.

From Case, Michelson went to the newly formed University of Chicago and spent most of his career there. Along the way, he carried out many ingenious experiments and repeatedly made attempts to increase his measurements of the speed of light. He was the first American citizen to win a Nobel Prize. It was awarded in 1907. As the incident concerning young Michelson's visit to President Grant indicates, he always tended to be self-centered to a significant degree and guarded carefully what he looked upon as his rights and privileges. This often made him somewhat prickly in his relations with others. Various stories about his actions at the Academy attest to this.

There is one story about Michelson's relations to others that bears repeating. While at the University of Chicago, he wrote a small book with the title *Studies in Optics* published by the University of Chicago Press in 1927. In it he displayed some of the wonderful things he had done. The book was not only well written but was well illustrated with photographs that were prepared by a very talented graduate student named John C. Clark, who held a research assistantship and was later a prominent leader of assembly teams at Los Alamos. Michelson was pleased with the quality of his work. When, however, Clark purchased a copy of the book at the university bookstore and asked Michelson if he would autograph it the great man refused.

THOMAS HUNT MORGAN

When Michelson retired from the presidency of the Academy in 1927, the brilliant geneticist Thomas Hunt Morgan (1866–1945) seemed to be a very reasonable candidate to replace him. He was broadly based in the biological sciences and had gained fame through a study of the genetic properties of the fruit fly with great emphasis on the details linking phenotype to the variance in chromosomal structure.

Figure 21.   Thomas Hunt Morgan

Morgan had completed his undergraduate work at the University of Kentucky in natural history in 1886 and obtained a Ph.D. at Johns Hopkins University in 1890. He received a research and teaching position at Bryn Mawr College in 1891 but then moved on to Columbia University in 1904 where he remained until 1928. In the intervening period, he spent a great deal of time at Woods Hole during the summers working very broadly in the biological world. However, he became more and more specialized on research dealing with the chromosomes of the fruit fly, rediscovering as it were an inner cell hereditary link that coupled closely with Gregor Mendel's earlier discoveries in plant genetics. The work on fruit flies apparently began in the period 1908–1909.

Along the way, he spent a good deal of time at the zoological station in Naples. The atmosphere there was international and provided Morgan with excellent opportunities to become familiar with many European biologists. His professional life became very rich.

The selection of Morgan as president of the Academy seemed in keeping with what might be called unwritten rules requiring that the president be in relatively close proximity to Washington. By 1927 Washington was only a four or five hour train trip away from New York. The California Institute of Technology, however, made Morgan a very attractive offer to head the division of biological sciences there in 1928 when he was sixty-two years old. After some hesitation he accepted the offer and with it inevitably changed his lifestyle. Air travel was essentially unknown at that time and, for the most part, the truly rapid so-called streamlined trains lay somewhat in the future. As a result Morgan could only maintain modest personal communication with the Academy, mainly by telephone and through intermediaries. His remaining years in office were not highly distinguished by much in the way of novel activity.

The Great Depression

To cap it all, the Great Depression struck the country in the autumn of 1929 and was not to lose its grip until the onset of World War II ten years later. The request for services from the National Research Council had actually dropped off rapidly once World War I war ended and the budgets of the federal agencies more or less resumed peacetime levels. The depression, however, exacerbated severely what was already a difficult situation. It was also aggravated by the fact that the charter of the Academy seemed to imply to some that it could only respond to requests and should not help in generating them unless asked, as was the case for President Marsh.

WILLIAM WALLACE CAMPBELL

The next president was the astronomer William Wallace Campbell (1862–1938), who held office between 1931 and 1935. He entered the University of Michigan as an undergraduate with the intention of majoring in some form of engineering but shifted his interest to astronomy as a result of attending some very inspiring lectures on the subject. He obtained a bachelor's degree in 1886. He had much mathematical ability and spent the following two years at the University of Colorado as an instructor in mathematics.

Figure 22.   William W. Campbell

In the meantime, the Lick Observatory in California had become operable (1888). It had been built with funds provided by James Lick (1796–1876), the wealthy California pioneer. Campbell began spending summers at the Observatory, joining its fulltime staff in 1891. He had several specialties. He made a practice of photographing the position of stars during total eclipses and used deviations to confirm the predictions of the general theory of relativity. He spent some thirty years measuring the radial velocity of some 1,000 stars, a tedious process, thereby establishing the spatial drift of the solar system.

Campbell was made president of the University of California at Berkeley in 1923 and was elected president of the Academy in 1931 at a time he began experiencing bad health that affected his eyesight. To simplify his work with the Academy, he gave up the presidency of the University of California. He and his wife moved to Washington where he remained until 1935. Being severely depressed by his ever-failing eyesight, he decided to take his life in 1938.

The relationship between the government and the Academy had reached sufficiently low ebb in the mid-1930s that Campbell had established an Academy committee carrying the designation Government Relations and Science Advisory Committee that attempted to stimulate activities of mutual interest. Unfortunately it was not very successful, in keeping with the dismal temper of the times.

James Lick

James Lick deserves mention. He was a Pennsylvania Dutchman who apprenticed in a factory in Baltimore that manufactured pianos. His family name was Lueck, which he simplified for practical reasons. In the course of events, he noted that not infrequently a dray would carry a number of pianos to the loading docks. On inquiring about the destination of the shipments, he learned that it was South America, which lacked a manufacturer. As soon as he had mastered the trade, he set off for South America and created several factories. It was a very dangerous time of revolution and crime, but he rapidly learned to deal with the turbulence by creating a small private army of his own to protect him and his properties.

He was in Lima, Peru, when the United States acquired California in 1848. He sold all his factories for gold and arrived in San Francisco with about $30,000 in hand, just before the discovery of gold in California. In the process he persuaded a local manufacturer of chocolate candy in Lima named Ghirardelli, a very good friend, to join him in going to California. On arriving there Lick promptly purchased all the real estate his money could buy, presumably much from fleeing Mexican landowners. His experience in South America had

Figure 23.   James Lick

taught him how to fend off squatters and thieves so he was soon one of the wealthiest individuals in California. Along the way he became known as "the generous miser" since his public-oriented charities were numerous and well appreciated by the community. Most survive to the present time.

Toward the end of his life Lick wondered what to do with the remainder of his substantial wealth after taking care of his existing charities and the security of special relatives who had joined him. There is a legend that he considered erecting a pyramid rivaling those in Egypt, but this seems out of keeping with his rock-bottom practical nature, as well as the potentiality of even his riches. Finally, he decided to use his remaining fortune to support the construction of the largest refracting telescope in the world at the time, twice the diameter of any existing one, and place it on Mount Hamilton, near San Jose, California. He also planned to be buried under it. And that is what happened. An announcement of the decision was made in 1873.

Lick had been only very marginally interested in science during his lifetime so there has been a reasonable amount of speculation about the factors that finally influenced him to create what became the Lick Observatory, now attached to the University of California. The historical records show that Joseph Henry first made the proposal. It is also possible that Benjamin Gould added stimulus during a visit the latter made to California after returning from his great research program in Argentina, to be discussed below. There is no doubt, however that a major factor in the process was his close friendship with Professor George Davidson, mentioned earlier, another quite remarkable member of the Academy who used astronomical telescopes both for pleasurable star gazing and as a working tool.

George Davidson

George Davidson (1825–1911) was born in England into a seafaring family of Scots. His family moved to Philadelphia when he was seven years old, and his formal basic education occurred there, partly in the hands of his family. He did, however, enter the Central High School of Philadelphia where he graduated at the head of his class in 1845. Alexander Bache, who was principal of the school for a portion of the period, noted this exceptionally capable youth and employed him part time to observe meteors and record their arrival. When Davidson graduated, Bache took him on as a junior member of the staff of the Coastal Survey to continue observational work.

In 1849, Davidson was transferred to the California Hydrological Survey in the newly acquired western lands then under naval command. They studied and surveyed San Francisco Bay and the Sacramento River and issued a report useful for navigation in 1850. They then continued a survey of the foggy northern coast and, along with the production of useful basic maps of hazardous areas, gave recommendations for the positioning of lighthouses. They were faced with hostile Indian tribes in the course of this work and, while dealing with their normal physical problems, were required to employ arms on occasion. In 1854, after completing the activity along the northern coast and extending it to Monterey and the South, they returned East to prepare a very useful directory of the Pacific Coast that became available in 1857. About this time, the possibility that the Southern States might start a civil war loomed on the horizon. Davidson, who had become a close confident of Bache, was shifted to secret work involving the planning of matters

Figure 24.   George Davidson

such as the defense of Philadelphia and the Delaware River and the security of Florida.

In 1867, following the Civil War, Davidson was sent to Panama to explore the feasibility of constructing a canal there. He did not encourage immediate action. Soon thereafter the possibility of purchasing Alaska from Russia arose, and he was asked to lead a group to appraise the situation, a task for which he was ideally suited. He had learned over the years that it was generally useful to cultivate the friendship of local Indians in the course of such explorations. In this case, that practice proved to be especially beneficial because he soon succeeded in forming a friendship with a Chief who knew many essential details about Alaska and provided him with valuable information not otherwise available, even from the Russians. Davidson supported the purchase.

In 1868, Davidson was forty-three years old. A decision was made to place him permanently in charge of the Coastal Survey in California, a position he retained until his retirement in 1895. He rapidly became the most prominent scientist in the state. He received an honorary appointment as professor of astronomy and geodesy at the University of California and lectured there regularly. He soon was a much-admired celebrity in San Francisco. The highest hill in that hilly city was named after him (Mount Davidson). Beyond this and receiving many honorary awards, both national and international, he represented our country on a number of international commissions in his areas of expertise. Seamen revered him for what he had done to make voyages along the coast safer. His general popularity was further enhanced when he set up a 6.4 inch refracting telescope for public viewing in a local park and supported it with accompanying lectures in astronomy. He formed a warm friendship with James Lick who left the issue of the planning and construction of the Lick Observatory in his hands. Lick particularly admired him as an individual who would devote his life to public service and did not focus on the accumulation of wealth.

It is important to add that Davidson developed two major baselines for mapping California with the use of triangulation. One was placed in Yolo County northeast of San Francisco. It was about 175 kilometers long and estimated to be accurate to within less than a centimeter. The other, of comparable accuracy, was situated in the vicinity of Los Angeles. Among many other things, he used the first to locate a point on the top of nearby Mount Tamalpais. About ten years later, an individual who had made an independent measurement of the location of the point on the mountain claimed that the asserted accuracy of Davidson's baseline was faulty. When Davidson was asked about the counter claim, he said: "My statements about the accuracy of my baseline are right. Mount Tamalpais has just managed to move a bit." The

Figure 25.   Benjamin Gould

great earthquake of 1906 made it clear that the Coast Range is measurably mobile.

Benjamin Gould

Benjamin Gould (1824–1896), mentioned above, was a New Englander, the leading American astronomer of his day and a founding member of the Academy. He had started his career by using telegraph signals to determine values of longitude at various positions on earth while employed by the Coastal Survey. He also pioneered the use of photography in the positioning of stars. His greatest work was to develop a stellar catalogue in South America using Cordoba in Argentina as a base. He had spent time at Goettingen University where he was offered the chair that had been occupied by Gauss, but he turned it down.

FRANK RATTRAY LILLIE

The next president of the Academy was the Canadian-born scientist Frank Rattray Lillie (1870–1947) who served between 1935 and 1939. He had attended the University of Toronto with the intention of aiming for the ministry but became enchanted with the natural sciences on being exposed to a summer at Woods Hole. As a result he took a temporary teaching period at Clark University and then attended graduate school at the University of Chicago, obtaining a degree in zoology in 1894. This was followed by a position at the

Figure 26.   Frank R. Lillie

University of Michigan (1894–1899) where he was an instructor in zoology and then by a full professorship at Vassar College, which he held for three years. He subsequently accepted a position as assistant professor of embryology at the University of Chicago. He was elevated to a full professorship in 1906 and became chairman of the department of zoology in 1910.

During his term as president of the Academy, Lillie took special interest in the National Research Council fellowships, making certain that the choices placed emphasis on quality. He also stimulated the creation of the Woods Hole Oceanographic Institute, established in 1938. He gave special attention to the latter on the belief that the country needed a number of such institutes and definitely needed one on the East Coast.

FRANK BARTON JEWETT

The next president, Frank B. Jewett (1879–1949), chosen in 1939, was a very fortunate selection. He had just retired as president of the Bell Telephone Laboratories where he had been an eminent leader and had recently moved to Washington as a trustee of the Carnegie Institution and advisor to President Roosevelt at a time there was great need in the Capitol for individuals with his form of experience.

He had grown up in Pasadena, California, at the time it consisted of only a few houses, a one-room elementary school and many orchards. He attended the secondary and college-grade schools provided by the nearby Throop Institute of Technology, later the California Institute of Technology. He

Figure 27.   Frank B. Jewett

graduated in 1898. It was his plan to follow in his father's footsteps and continue his education at the Massachusetts Institute of Technology. Unfortunately, he was out of phase with the application cycle there. On the suggestion of a friend, he decided to undertake graduate work at the newly opened University of Chicago where Michelson agreed to take him as a student. He obtained a doctor's degree in 1902, leading to an instructorship at MIT.

While in Boston, he met by chance a scout for the American Telephone and Telegraph Company (AT&T) who was looking for new young talent to provide leadership for the task of extending the range of the telephone system. At that time, the New York based system had been marginally extended as far as Chicago, using primitive repeaters that employed relays consisting in the main of amplifying telephones that gave very poor performance because of the accumulation of background noise. The chief executive officer of AT&T, Theodore Vail, had insisted that it was time to develop a truly workable telephone network that extended all the way to San Francisco. Realizing that new technical talent would be needed, Vail sent out scouts to search for gifted individuals. One of them visited MIT and had a chance meeting with Jewett. The scout was much impressed by the combination of characteristics of this young man, as were many other individuals in the course of Jewett's life who had the good fortune to know him. He was not only highly knowledgeable but also had the qualities of leadership that came first in the search. Jewett was at that time tied up with a contract at MIT, but the AT&T scout came back when it expired in 1904 and convinced Jewett to join the engineering department of AT&T.

During his days at the University of Chicago, Jewett had become a good friend of Robert A. Millikan and arranged a meeting with him to discuss the problems he faced. Millikan recommended that he turn to the newly opened field of electronics, which he felt must contain a solution to the problem at hand. He also mentioned that Lee De Forest had developed some triode vacuum tubes which, although primitive, seemed to exhibit the type of amplifying properties AT&T was after. Jewett searched through the group of young individuals emerging from the University of Chicago for a helpmate and selected H. D. Arnold, who joined him in a diligent time-consuming search for a solution to their problem. Arnold carried on the detailed day-to-day investigations. The company purchased the patents of De Forest's triode and began to improve upon it in a very systematic way. Arnold is credited with the successful transformation of De Forest's very primitive triode into a highly reliable device. Jewett once said that during the uncertain period of development he and his colleagues got little sleep.

It so happens that the General Electric Company had decided independently that an electronic triode vacuum tube was the solution to many problems in communication technology and had begun serious investigations in the field under the leadership of Irving Langmuir. A lawsuit developed which AT&T won in principle. However, the major companies in vacuum tube development eventually agreed to create a patent pool, The Radio Corporation of America (RCA), at the end of World War I after deciding that they should not waste time and fortune quarreling with one another.

The Western Electric Company was the manufacturing arm of the telephone company and possessed a research and development component, designated its engineering department, which suited its immediate needs. Jewett became its assistant chief engineer in 1912, but rose through ranks to become president and director in 1921.

In 1925, the AT&T engineering department that had developed the successful vacuum triodes was combined with the engineering department of Western Electric and incorporated as the Bell Telephone Laboratories. Jewett was selected to be its president. It rapidly became one of the most distinguished laboratories in the world. Its great alteration and contraction in recent years, almost equivalent to demise, is little short of tragic. Isadore Rabi, a New Yorker and a keen observer, once made the comment: "The people they hire all look alike!" This statement probably reflected the evolution of a common, well-knit culture.

Jewett retired as president of the Laboratories within a year after being elected president of the Academy, and moved to Washington where he lived an active second life. Along the way he had become one of the leading figures in the evolution of the telephone company. Part of his success lay in the special form of charisma he exuded in his working relations with others. No one needed to fear approaching him to discuss a bothersome problem. He would always respond in a cheerful, constructive way. Once he arrived in Washington he was quickly drawn into the special White House circle that President Roosevelt put together to provide scientific leadership in World War II.

Leslie R. Groves

One example of Jewett's special form of personal service relates to the advice given to General Leslie R. Groves when the latter was appointed military head of the nuclear bomb project in 1942. Groves, who was a member of the Army Engineering Corps, had just completed the construction of the Pentagon building when the decision to go ahead at full speed with the bomb project was made. The White House group decided that Groves would be an ideal candidate for the new military post since he was greatly experienced in problems related to supply and construction. He, however, had grave misgivings about the situation, since he had no experience in dealing with scientists and wondered how to get along with these strange individuals. One of his colleagues suggested that he discuss the situation with Frank Jewett, who had such experience and an office nearby at the Academy. Jewett not only put Groves at ease, but also became a friendly advisor. General Groves in fact played out his role exceedingly well, making certain that needed supplies were always on hand to the extent humanly possible and that basic schedules of construction were met. Jewett had helped bring out Groves' strongest qualities. Richard Tolman, a distinguished scientist-administrator from the California Institute of Technology, eventually became Groves' official science advisor during most of the war.

Jewett had obviously come into office with a highly diversified and productive career behind him. In addition he was profoundly thoughtful in approaching a new situation. In reading the charter that created the Academy he concluded that it was a remarkable document, ingeniously contrived by its authors. He decided that at its core it was intended to be highly permissive as far as the professional activities of the Academy are concerned. The restrictive items related principally to practical money matters such as reimbursement for out of pocket expenditure accrued in the course of its legitimate activities. He also decided that the phrase "on request" did not imply that the Academy could not seek tasks from the government agencies, but rather that there should be mutual agreement that the Academy could take them on.

Jewett was also bothered by the fact that the offices of the president of the Academy and that of chairman of the National Research Council were separate. Moreover, one had a part-time and the other a full-time chairman. The arrangement had worked well over the years only because the two individuals had been close friends. In another situation it could lead to disaster. He decided to leave the resolution of this potentially explosive problem to his successors.

When Franklin Roosevelt became president in 1932, he and the Congress had initiated such organizations as the Public Works Administration to help create employment. Some members of the scientific community saw in this an opportunity to obtain governmental support for scientific research, basing their hope on the thought that new discoveries in science would have a stimulating effect on the economy. At least two serious proposals of this type were made as a result of activities within the Academy. President Roosevelt took a personal interest in them, but the members of his cabinet concerned with such plans vetoed them since they did not stimulate many new jobs in proportion to the money invested. Moreover, many members of the Academy feared that the formation of what could become deep links with the government in the process of determining research policy might result in the politicization of science and become counterproductive. As a result the issue of obtaining broad governmental support for basic science became a hopelessly controversial topic within the Academy and died in the process.

The situation began to change radically when World War II broke out in September of 1939. By that time many of the academic scientists, as well as others, in Great Britain had been mobilized into governmentally supported organizations and had in effect "gone underground". Special emphasis there focused on radar and fighter planes, initially for defensive purposes. Some attention was given to recently discovered nuclear fission, but the British were prepared to farm such studies out to available trustworthy scientists in the

Figure 28.   Vannevar Bush

United States and Canada since the topic was looked upon as being of secondary importance in the initial stages of the war.

In the meantime, the federally funded laboratories in United States that had relevance to national defense began to receive expanded budgets and became focused on topics that were likely to be immediately useful in the war.

The American scientific community was keenly aware of the changing situation in Europe and began to establish professional committees that individually focused on some broad area of military-oriented research, drawing upon the talents and interests of its members. To the extent possible, foundations or individual philanthropists funded the work of such committees initially. One might have thought that the National Research Council would have played a major role in the organization of the scientific community for military research as it did in World War I, but this did not occur. It was clear from the start that the funding of such research would eventually be shifted to governmental sources. Apparently the disputes within the Academy regarding the advisability of using federal funds for research that had taken place during the 1930s made it seem advisable to try new approaches worthy of the magnitude and urgency of the developing international crisis. It is said that Winston Churchill had emphasized to President Roosevelt the great importance of involving the scientific and technically oriented academic community in the heart of the defense structure. It had proven to be invaluable in Britain.

National Defense Research Committee

The first step in providing a new solution to the use of science was taken by President Roosevelt in June of 1940, immediately following the fall of France. It led to the formation of the National Defense Research Committee (NDRC) in the Executive offices of the government and close to the White House. Its first chairman was Vannevar Bush (1890–1974), president of the Carnegie Institution in Washington since 1939 and previously at the Massachusetts Institute of Technology. This was followed in May of 1941 by the creation of the Office of Scientific Research and Development (OSRD), again in the Executive structure. It had the ability to fund large-scale research programs, including those at universities. Bush became its director and James B. Conant, the president of Harvard University, became the Chairman of NDRC, which was automatically attached to OSRD.

James Conant

Conant (1893–1978) had centered his academic career about Harvard where he established a world-renowned laboratory in organic chemistry after

Figure 29. James B. Conant. Photo by Karsh of Ottawa.

completing his basic and advanced education there. He retained close ties with his American and European counterparts, but was conscious of the great dangers that Hitler posed once the latter appeared prominently on the political scene in Germany. He not only gave early warnings about this but broke off all possible connections with previous German colleagues and friends who supported the dictator. He possessed so many outstanding qualities that he was chosen to be the president of Harvard University in 1933 and held the position for twenty years with a leave of absence for wartime duties.

In 1946 President Truman offered Conant the post of chairman of the newly-formed Atomic Energy Commission but he was very anxious to return to the responsibilities at Harvard. He did however accept membership both on the General Advisory Committee of the Commission chaired by Robert Oppenheimer and the chairmanship of the National Science Board, which oversaw the activities of the newly formed National Science Foundation. Then, in 1953, he resigned his post at Harvard while accepting an invitation from President Eisenhower to become the first High Commissioner to a Germany that was in the process of being reformed. He remained in the chair for four years as a very important and popular figure in that important transitional stage. He presided over much of the activity that accompanied the implementation of the treaty that created the Federal Republic of Germany. Throughout most of his career, Conant devoted a great deal of time to discussions and writings on principles that should govern educational processes.

Frank Jewett was not a member of the NDRC, but could attend the meetings as somewhat in the nature of a private advisor to President Roosevelt as well as president of the Academy. He rapidly became an effective "insider". It was agreed that the Academy would cooperate in all possible ways with the OSRD. In fact, several committees of the NDRC used the facilities of the Academy for their headquarters. This included the committees on medicine and on ordnance. As a result, the Academy building soon became cluttered with temporary wooden partitions.

When the war ended with the collapse of Japan in the late summer of 1945, the role of the Academy began to change in a much more positive direction. Two factors were involved in major ways. First, the OSRD and the NDRC were disbanded (1947) and the various executive agencies were given freedom within budgetary restrictions to execute their own research programs. By this time, so many academic scientists had been supported by the OSRD during the war that the objections to receiving support for scientific research from the government no longer seemed valid.

Vannevar Bush and Karl T. Compton, the president of MIT, took posts within the Department of Defense in a desire to provide special guidance there on technical and scientific matters. They headed in turn an office designated the Defense Research and Development Board (DRDB) patterned somewhat after the wartime OSRD. However, the now highly experienced staff in the Pentagon wished for more freedom of action. A compromise was ultimately reached by the creation of the Defense Science Board (DSB), a review and advisory body that has endured since. The president of the Academy was designated as a statutory member of the Board, although not all presidents have exercised the privilege of participation.

The new order included the extensive transformation of some old science-based agencies, such as the National Institutes of Health (NIH), or the creation of entirely new ones, such as the National Science Foundation (NSF) and the Atomic Energy Commission (AEC). Also existing agencies added special scientific adjuncts, such as the Office of Naval Research (ONR).

The first attempt to create a science foundation that would support the basic sciences failed. The group of scientists who helped to prepare the supporting legislation proposed a plan whereby all decisions regarding the disposition of research funds would be determined by a board of scientists chosen by the scientific community. The government would merely determine the allotment of funds. President Truman found it necessary to veto the legislation. It took a year or two to straighten matters. In the meantime the AEC and the military science agencies, such as ONR, supported much research.

Many of the new organizations as well as older ones began to ask the Academy to form advisory committees within the National Research Council that could be of help in their various fields of interest. The Academy's advisory structure began to take on new life.

Detlev W. Bronk

The second major post-war transformation at the Academy involved the arrival on the scene of a remarkably versatile scientist, namely Detlev W. Bronk (1897–1975). His undergraduate education in electrical engineering at Swarthmore College was interrupted by World War I service in training for naval aviation. He started graduate studies and research in the physics department of the University of Michigan with initial emphasis on investigations in the infrared portion of the optical spectrum, one of the major areas of research in the department at the time. In the process, however, he became interested in the physics and physiology of nerve conduction and finally obtained a doctor's degree in 1926 in the combined disciplines of physics and physiology. He then received joint appointments at Swarthmore and the medical school of the University of Pennsylvania, carrying out teaching at one and research at the other.

This period was interrupted for a year (1928–1929) on an NRC fellowship that he spent in Cambridge, England, working with Professor E. D. Adrian, F.R.S., who was carrying on very advanced research in the field of nerve conduction. The two became intimate life-long friends. Adrian eventually shared a Nobel Prize with an English colleague for his research.

On returning, Bronk not only received a professorial appointment at Swarthmore, but was made the director of the newly established Eldridge Reeves Johnson Foundation for Medical Physics at the medical school of the University of Pennsylvania. The Johnson fortune had been partly earned as a result of the commercial success of the once celebrated Red Seal recordings of classical music marketed by RCA.

Figure 30.   Detlev W. Bronk

Bronk's principal service in World War II had been as Coordinator of Research in the Office of the Army Air Surgeon where he displayed his usual dynamic and imaginative leadership.

In 1946, President Jewett of the Academy, having come to appreciate the full range of opportunities that the institution offered in the immediate post-war world, asked Bronk if he would be willing to accept the post of full-time chairman of the National Research Council. Bronk agreed with the understanding that he owed residual responsibilities to the University of Pennsylvania. During these discussions, both Jewett and he apparently also agreed that the principle the Academy had in the main followed since its creation in not seeking advisory activities for the National Research Council unless requested to do so by a federal agency had been interpreted too narrowly. The post-war period required a more liberal and exploitative view of the matter that made full use of the national scientific talent.

In 1948 Bronk added the position of the presidency of Johns Hopkins University to his growing list of responsibilities. Then in 1953 he switched to the position of president of The Rockefeller Institute in New York, a post that allowed more flexibility.

Bronk soon took charge of most affairs at the Academy in his highly creative way. On one occasion, at a small gathering of members, someone commented that Bronk was now effectively chairman of almost all relevant Academy committees. Another member of the group jokingly suggested that they should form a committee in which Bronk was explicitly excluded. A third responded: "It would be futile; Bronk would be its chairman within a week after it was formed!"

On retirement from his chair in 1947, Jewett gave a brilliant address in which among other things he outlined his personal views of the future potentialities of the Academy as it advances. The address was reprinted in the proceedings of the Academy, *PNAS* 48(1962): 481–90.

ALFRED NEWTON RICHARDS

Jewett was followed as president in 1947 by a research physiologist, Alfred Newton Richards (1876–1966), a close friend of Bronk and also from the University of Pennsylvania. He was quite willing to give Bronk all the leeway he needed to rejuvenate the Academy as a major advisor to the government on scientific matters. This meant sitting on the sidelines while Bronk displayed his typical inspired activity.

Richards had a particularly distinguished career. His father was an impecunious clergyman who sacrificed sufficiently to send him to Yale as an

Figure 31.   Alfred N. Richards

undergraduate. There he was much stimulated by an undergraduate course in chemistry given by Professor R. H. Chittenden. The latter noted his interest and suggested that he continue in graduate work in chemistry. When Chittenden learned of the financial limitations that governed Richards' actions, he arranged a very modest fellowship from Yale that carried Richards through the first graduate year. Fortunately Chittenden obtained an auxiliary appointment at Columbia University and was able to shift Richards to it along with a sequence of better-paying appointments that eventually extended to a professorship in pharmacology in 1903.

Along the way, Richards greatly broadened his knowledge and laboratory skills in several fields. In 1908, he accepted an appointment in the medical school of Northwestern University in Illinois where he began to study the detailed action of drugs. He made the close friendship of Christian Herter and formed links with the newly created Rockefeller Institute for Medical Research in New York. In 1910 he shifted to the medical school of the University of Pennsylvania, which served as his main base until he became emeritus.

He joined the OSRD as head of the Committee on Medical Research in 1941, centering its main offices in the Academy building at 21st and Constitution Avenue. The committee worked in many areas including treatments for tropical diseases and blood substitutes.

When Professor Howard Florey, F.R.S., arrived from England during the war seeking advice on ways to enhance the production of penicillin, which was miniscule at the time, he was faced with a conflict between those who wanted to continue by enhancing production through fermentation and those who preferred experiments on synthesis. Richards strongly advised the first route and put Florey in touch with the Department of Agriculture's Research Laboratory in Peoria, Illinois, where large scale fermentation procedures were being developed. Major pharmaceutical corporations took over from there. His advice proved to be on the mark. Copious quantities of the antibiotic were available by the time of the Normandy invasion in June of 1944.

DETLEV WULF BRONK

Finally, in 1950, Richards decided that Bronk was sufficiently deeply involved in a major transformation of the Academy that the latter should appropriately replace him as president. The change took place in 1950. Bronk held the post for twelve years but continued as president of The Rockefeller Institute for Medical Research in New York, which under his leadership became The Rockefeller University. Meanwhile the goals he and Jewett had envisioned for the Academy had been achieved. Many well-established conventions were overturned as the range of the Academy broadened. In effect the Academy was now administered by what amounted to a full-time president who held the key posts of governance.

When Richards gave up the presidential chair, a formal nominating committee selected by the Council of the Academy had designated James Conant as the replacement candidate with the understanding that it was to remain a part-time position. Many of the members of the Academy objected not to Conant but to a continuation of the principle that the post of president be part time with the connotation that it was substantially honorific. As a result, a group of scientists who were intimately aware of Bronk's formidable dedication to planning a new future for the Academy placed him in nomination. Once Conant understood the issues involved, he withdrew his own nomination. It might be added that when Bronk was offered the post at The Rockefeller Institute, he mused for a bit and then said: "Yes, I think I could handle that too", implying that he perhaps intended to retain most or all of his other prestigious but burdensome administrative posts. The Rockefeller board made it clear that such a plan was not part of the deal.

As mentioned, up to this time the chairmanship of the National Research Council was not normally held by the President. Bronk soon decided after taking on his new office that the two positions should be joined to achieve optimum effectiveness in the management of the Academy. They were combined with the approval of the Council.

The Battery Additive Crisis

A special incident that eventually attracted some nation-wide attention occurred in 1953. A commercial organization began marketing an additive that it claimed would enhance the life of ordinary wet-cell lead-plate batteries. The National Bureau of Standards, headed by an excellent scientist, Dr. Allen V. Astin, carried out a routine study of the additive and determined that it was not effective as claimed. The Bureau ruled that it should be taken off the market. The head of the company was a good personal friend of a top-ranking member of the staff of the Department of Commerce, a new appointee. The latter presumably was not familiar with the ways of governmental actions and made the mistake of attempting to fire Astin. Bronk joined with Dr. Mervin J. Kelly, the president of the Bell Telephone Laboratories and chairman of the scientific advisory committee to the Department of Commerce, to rectify the situation, which had Washington in an uproar. Their intervention led to the reinstatement of Astin.

Mervin Kelly

It should be added that Kelly (1874–1971) was a very influential figure on the scientific scene at this time, being a highly placed advisor to many prominent organizations as well as the government. He had a strong character and a lively imagination. Born in Missouri, he obtained a doctor's degree in physics under Robert Millikan at the University of Chicago. He then joined AT&T and was assigned to the laboratories of The Western Electric Company where

Figure 32.   Mervin J. Kelly

Figure 33.   Patrick E. Haggerty (left)

he was in charge of improving the power and other operational properties of the vacuum tubes employed as relays in the telephone system. He was very successful in pushing that form of technology to its limits.

In the 1930s, he became fascinated with the properties of semiconductors, which were naturally conducting at ambient temperatures and had been used as rectifiers and heterodyne mixers since the turn of the century. He decided that it should be possible to develop a triode replacement for the vacuum tube with them in some way. A number of individuals had tried and failed and were continuing to do so. His opportunity to pursue this goal came in 1936 when he was appointed Director of Research of the Bell Telephone Laboratories. He gave the assignment of inventing such a triode to Dr. Walter H. Brattain one of the existing staff and Dr. William S. Shockley a new employee. The war intervened but fortunately led to the exploitation of elemental silicon and germanium as rectifiers for radar, providing excellent new candidates for experimentation after the war. Kelly reopened the quest, forming an enlarged team with the addition of John Bardeen and others with well-known success. Kelly can be regarded as the spiritual father of the transistor. His wisdom and vision was displayed by his decision to license the rights to produce the transistor to appropriate companies for the trivial price of $25,000 in order to accelerate development.

Texas Instruments

Kelly balked when he first received a request from Texas Instruments. It was a company that had been engaged in searches for likely oil-bearing strata by acoustical means before the war, but had been aided by an extraordinary young naval-volunteer engineering officer, Patrick E. Haggerty, in developing magnetic detectors for use in the search for submarines during the war. Haggerty had joined the company and was actively involved in a complete reorganization. Once Kelly met Haggerty, he realized that he had encountered a visionary mind comparable to his own and relented. Texas Instruments took the lead in transistor development for the next quarter of a century. Kelly is shown in Figure 33 chatting with David Rockefeller on the campus of The Rockefeller University.

The Trial of J. Robert Oppenheimer

One of the tragic events that marked Bronk's early period in office concerned the trial of Robert Oppenheimer by the U. S. Government in 1954. The trial stemmed from concerns about his ultimate loyalty to that government. Oppenheimer had served brilliantly as the director of the Los Alamos Laboratory during World War II and justly deserved the many honors he received for his extraordinary leadership, including the position as chairman of the General Advisory Committee to the Atomic Energy Commission. When, however, President Truman proposed an exploratory program on the development of fusion weapons involving light elements as a safeguard against prior development by the Soviet Union, which had successfully produced a fission bomb in only four years, Oppenheimer, speaking for himself and his committee, expressed opposition. An investigation disclosed a not uncommon left-wing past that in his case was deemed to be sufficiently significant to revoke his access to classified material.

Harrison Brown

The activities of the foreign secretary of the Academy had traditionally been confined almost exclusively to relationships with the academies of Western Europe. When the office became vacant in 1961, Bronk appointed Professor Harrison Brown (1917–1986), a geochemist at the California Institute of Technology, to the post in 1962. The latter had wide-ranging ambitions to form bonds with science academies throughout the world, including those in communist lands, and achieved a remarkable degree of success in the process. Unfortunately, Brown became incapacitated prematurely through ill health. He did however succeed in establishing a new, greatly extended pattern for the office. Bronk had come to know Brown well as a result of his stimulating chairmanship of the Academy's Committee on Oceanography (1957–1962).

One of Brown's great ambitions was to establish relationships with the Mongolian Academy of Sciences in Ulaanbaatar when the province was under Soviet domination. Unfortunately time ran out.

The appointment of Brown, who had constructive ideas of the manner in which an expanded Office of the Foreign Secretary might better serve the promotion of international relations in science, demonstrates the way in which Bronk encouraged members of the Academy to display initiative in promoting scientific enterprises. Many other examples could be cited.

The IGY: Lloyd Berkner

One other case that deserves special mention is represented by the evolution of the program involved in the International Geophysical Year (IGY) (July 1, 1957–December 31, 1958). On two previous occasions, the international scientific community had joined forces to carry out cooperative studies of the polar regions, namely in 1882–1883 and 1932–1933. One of the distinguishing events in the IGY that took place in the early 1930s was the introduction of the first microsecond pulsed radar by Dr. Hans Hollmann who used it to study the ionosphere in the polar region, but also noted that he obtained good images of the surrounding terrain on his oscilloscope.

In 1950 Dr. Lloyd Berkner (1905–1967), who was associated with the Carnegie Institution in Washington at the time and who was widely engaged in promoting new scientific activities and institutions both at home and abroad, proposed that the situation was ripe for the development of a worldwide program of eighteen months duration that would be devoted to geo-

Figure 35.   Lloyd V. Berkner

physical research in all frontier areas. While most of the funding for U. S. participation came from the National Science Foundation, Bronk provided full encouragement and cooperation through the Academy.

Berkner, incidentally, had grown up during the period in which wireless and vacuum tube technology had matured and was a master of the field in its applications to science and engineering. His research at the Carnegie Institution had focused on gaining improved understanding of the ionosphere and its structure.

The seasonal appearance of an ozone minimum in the Antarctic was one of the many important discoveries made as a result of activities growing out of the IGY. One of the inventors of one of the techniques used, Gordon Dobson, F.R.S., proposed as a hypothesis, that it was a result of the appearance of annually periodic circumpolar cyclonic winds that prevent the influx of ozone generated in the tropics, a mainly physical phenomenon. Studies made since strongly indicate however that the destruction of ozone is primarily chemical in nature (See, for example, the review paper by Paul A. Newman, Antarctic Total Ozone In 1958, *Science* 164(1994): 543–46. Also major advances were made in understanding ocean-floor movement and continental drift.

Berkner furnished much of the initiative that led to the creation of the Arecibo radio telescope in Puerto Rico and the National Center for Atmospheric Research in Boulder, Colorado. Actually he displayed great persistence and imagination in pursuing important goals even as a teen-ager. One day in the early 1920s, he noted in the local Milwaukee newspaper that the Air Force was planning to offer twenty individuals in his age group an opportunity to spend that summer at one of its air bases where they would gain first-hand experience with military aviation. Applications were welcome. Lloyd said to his closest friend: "You and I are going to apply!" His friend said: "It's open to the whole country. We don't stand a chance." Lloyd's response was: "Let's at least apply." Their applications were not accepted, whereupon his friend said: "Just as I thought." Lloyd responded: "Look, you and I are going down to that air base, sit in on the lectures and do everything that the other guys do, all on our own." The two proceeded to follow this plan. One day not long after they started the proposed regime and were following the calisthenic routine of the main group in the exercise field, but out to one side, the sergeant in charge shouted: "Hey you two guys get in line!" Berkner and his friend eventually became air admirals in the Naval Reserve. His distinguished military career had in the main been devoted to applications of forefront electronics to important practical problems. His peacetime work knew almost no limits.

As his final endeavor, Berkner moved to Dallas, Texas, and with a colleague, Lauriston C. Marshall, established a technical school that reflected his personal views of the proper approach to such education. On Berkner's death, the institution was absorbed into the state university system and is now the University of Texas at Dallas. Berkner was buried in Arlington National Cemetery.

In the course of their work on the new institution, Berkner and Marshall prepared a scientific paper in which they proposed that the pattern of evolution in the post-Cambrian period was greatly influenced by a large increase in the chlorophyll-produced oxygen content of the atmosphere. This speculation appears to have survived the test of time.

James E. Webb, the future administrator of the National Aeronautics and Space Administration (NASA) in the 1960s, came to know Berkner at the height of the latter's creative period and would often ask both while leading the Agency and later: "Where are the Lloyd Berkners of today?"

Centennial Celebration

Although retired, Bronk was elected by acclimation to serve as chairman of the Centennial Celebration Committee of the Academy when it attained its hundredth birthday in 1963. The celebration was lavish and memorable for the quality of the scientific lectures[5] that were presented during the event. In connection with it, the new president and the Academy Council commissioned a noted science historian, Rexmond C. Cochrane[6] of Johns Hopkins University, to prepare a history of the Academy's first century.

By the time Bronk left office in 1962, the number of active advisory committees working within the combined framework of the Academy and National Research Council numbered in the hundreds. The organization had become both lively and productive in a sense it had not achieved since World War I. It had again become an indispensable component of the national scientific edifice.

Along with his many other activities, Bronk had an opportunity to expand the Academy's headquarters on Constitution Avenue by the addition of conforming wings on the east and west sides of the main building, thereby creating centrally located offices for some of the divisions of the National Research Council. The West Wing was funded by a generous grant from the Equitable Life Assurance Company and by special agreement carries a plaque that indicates the source of the funds. The East Wing was financed by a collection of grants from private foundations and industry. Wallace Harrison, who had helped design the original building under Goodhue, was the architect for both wings.

In departing from his post, Bronk proffered advice to his successor regarding a possible means of consolidating much of the remaining activity of the National

Research Council in a single building. He had learned by chance that George Washington University had acquired much old real estate in its neighborhood and was looking for tenants who might construct and lease replacement buildings. The new president followed this suggestion, which ultimately led to what became the Joseph Henry Building at 21st and Pennsylvania Avenue. It was constructed with plans provided by the Academy and leased for twenty years from George Washington University. Other solutions for consolidation of the growing enterprise were found at a later time, including a newly constructed building at 500 Fifth Street, N.W., in downtown Washington that was opened in 2003. It was funded by the Academy, but named in honor of the Keck Foundation in recognition of a special grant, The Keck Initiative Program.

Sputnik; The Creation of NASA

In the meantime, the Soviet Union had launched a small earth-orbiting satellite, *Sputnik*, during the autumn of 1957. As a result, the population of the United States was overcome by something in the nature of frenzy since it appeared that Soviet rocket technology was now much ahead of that in United States. Actually our naval scientists had been given authority to develop a small satellite that was intended to be used only for scientific observations, but the program had been noted more for its failures than successes when *Sputnik* appeared in the sky. Fortunately, a group working under The Army Ballistic Missile Agency at Redstone Arsenal in Huntsville, Alabama, had long since prepared a back-up rocket, but had been constrained from launching it since such a step, taken by the Army, might have been regarded as "provocative" on the national as well international scene. In any case, a U. S. Army satellite launched from Cape Canaveral joined *Sputnik* three months later under the guidance of a group of scientists including Wernher von Braun, James A. Van Allen and William H. Pickering, much to the relief of the population.

One other consequence of the situation was the passage in Washington in 1958 of legislation needed to create a new federal agency to be called the National Aeronautics and Space Administration (NASA), along with the termination of the National Advisory Committee on Aeronautics (NACA). More will be said about this later.

The President's Scientific Advisory Committee

In part to ease public anxiety, President Eisenhower immediately appointed a high level President's Science Advisory Committee (PSAC) attached to his office that was to provide continuing advice on scientific matters both

Figure 36.   George B. Kistiakowsky

self-generated and on request. The first chairman was James Killian, the president of MIT, the second was George Kistiakowsky of the chemistry faculty of Harvard, the third was Professor Jerome Wiesner of MIT and the fourth was Professor Donald Hornig of Princeton University. When President Nixon took office in 1968 he appointed Dr. Lee A. DuBridge, the former president of the California Institute of Technology to the position for two years, with the condition requested by the new Secretary of Defense, former Senator Melvin Laird, that the committee cease reviewing the affairs of his department. Dr. Edward David a former member of the Bell Telephone Laboratories held it for the remaining two years of President Nixon's first term. The committee as originally conceived was eliminated early in 1973.

In response to the now rapidly developing situation in space science and technology the academy created a Space Science Board in 1958 with Lloyd Berkner as its chairman. Its purpose was manifold and in part to develop links with the members of the international scientific community interested in the field of space research.

When George Kistiakowski departed from his position as the chairman of the President's Scientific Advisory Committee in 1961, he approached Bronk with the proposal that the Academy create a counterpart committee that could work in tandem with PSAC to the extent possible and operate with more openness and freedom. Bronk, who was serving as a member of PSAC, was enthusiastic about the concept. When the plan was fully formulated and approved by the Academy's Council, it bore the name Committee on Science and Public Policy (COSPUP). Kistiakowski was its first chairman, bringing with him much of the experience he had gained at the White House.

FREDERICK SEITZ

The new president was Frederick Seitz (1911–), who had served on the Academy's Council previously and held the office between 1962 and 1969. He quickly realized that Bronk's initiatives had generated a large and continuing increase in the advisory activities of both the Academy and the National Research Council. It would no longer be possible to manage them on a part-time basis. On reviewing this matter, the Council of the Academy decided that henceforth the president's position would be full time, and for a maximum of two six-year terms. Bronk had resolved the problem of dealing with the increased responsibilities that go with growth in his own way, first by accepting the position of president of Johns Hopkins University in nearby Baltimore, an easy commute to and from Washington, and then by shifting to a less conventionally demanding post in New York.

The 1960s began to yield many of the fruits of the scientific endeavor that had begun during and after World War II following the discovery at The Rockefeller University in 1944 by Avery, McCarty and MacLeod that strands of DNA carry the genetic message, and the subsequent development by Watson and Crick of the entwined pair model for the genetic units that reside in the cell and ultimately determine the characteristics of the living organism. By the 1960s, many of the details of the precise operation of the genetic system, including the genetic code, were being revealed. Biology at the molecular level was now a very open field for research with promises for the development of entirely new areas of clinical medicine.

Figure 37. Frederick Seitz

Figure 38.   Maclyn McCarty M.D.

Professor Maclyn McCarty, M.D., the pioneering biochemist of the Avery team at The Rockefeller University (and much more), is shown in Figure 38. In 1983, he used the facilities of the Commonwealth Book Program to present a detailed account of the course of the discovery: *The Transforming Principle* (W. W. Norton).

Similarly, the studies of the composition of the atomic nucleus with accelerators of ever-higher energy revealed the existence of a plethora of new particles. A great deal of effort has gone into the process of establishing their interrelation with much initial success. In parallel with this were reviews of studies regarding the manner in which the heavier chemical elements are created from light ones by accretion in highly radioactive stars.

It had long been known that there were links between fossils in western Africa and eastern South America. In 1912, the Austrian geophysicist Alfred Wegener proposed that the continents are able to drift on a semi-plastic base and may occasionally be joined for periods of time. Most leading geologists did not accept Wegener's theory but, starting in the 1930s, Harry Hess (1906–1969) of Princeton University decided that the concept deserved serious testing and began measuring the ages of the basaltic rocks that form continuous platforms or "plates" on the ocean floor. His speculations at this early time were augmented by those of a remarkably prescient English geologist, Arthur Holmes.

Harry Hess: Roger Revelle

Hess joined the naval reserve and began carrying out measurements from naval vessels at every opportunity before, during and after World War II. His

Figure 39.   Harry H. Hess

very active wartime service began the day after the bombing of Pearl Harbor. The first assignment placed him as one of the leaders of a group involved in the successful search for German submarines off the Atlantic Coast. He then became commander of an attack transport in the Pacific Ocean for an extended period. This activity provided much opportunity for oceanic research in between active military engagements. He discovered that none of the rocks in the basaltic base of the ocean floor were older than one hundred million years, in support of the view that the floor is subject to a mechanism that produces flow and replacement of older material by subduction under continents or by the production of new land. The mechanism at work is presumably adequate to move continents and build mountains.

In 1960, Hess prepared a report proposing that volcanic activity along mid-ocean ridges plays a key role. The determination of the exact nature of the process involved presented a major geological puzzle that was finally resolved by studies carried out by F. J. Vine and D. H. Mathews, two English geophysicists, and by Tuzo Wilson, a Canadian. The first pair studied the orientation of the magnetic traces found on rocks that flow out on either side of the ridges. Such traces alternate their magnetic signature with reversal of the earth's magnetic field. The experiments showed that the traces are symmetrically disposed on either side of the fissures where the upwelling of magma occurs. The basic model for the system was completed by Wilson, who proposed that the forces responsible for ocean floor motion arise from convective currents beneath the ocean floor that exert tangential forces upon a system of relatively rigid basaltic plates that are generated by solidified magma more or less continuously exuded from volcanically active ocean ridges. The research involved in these revelations was just coming to fruition during the Academy's centennial celebrations and caused much excitement among those interested in the topic. Unfortunately, Hess who had zealously persisted in pursuing the matter from its unpopular beginnings died prematurely soon after. He was raised to the rank of rear admiral in the reserve as a result of his wartime service.

Hess had been asked to prepare an address on this rapidly unfolding topic at the centennial meeting of the Academy, but was much too involved in research at the moment to do so. His admiring friend Roger Revelle, an oceanographer of comparable stature, prepared Hess' contribution to the occasion, all under Hess' name.

It may be added that in summarizing Roger Revelle's status as a scientist and human being in his *Memoir*, three of his professional colleagues stated: "Roger Revelle was one of the twentieth century's most eminent scientists. His life's work personified Ernest Boyer's four categories of scholarship: discovery, integration, dissemination and application of knowledge. He brought his talents in these categories to bear on the study of the planet we inhabit and our interaction with that planet. His interests and intellectual reach spanned the physical, biological and social sciences, engineering and humanities. He enhanced the status of oceanography in world science, pioneered in the study of global warming and brought a fresh approach to issues of population, world poverty and hunger. Revelle was an inspiring leader of scientific enterprises and an insightful and sagacious educator. He was the intellectual architect for the creation of a great university. He excelled in the communica-

Figure 40.   Roger Revelle

tion of science and its implication to policy makers and to the public. Revelle was an exemplary citizen in his community, his country and to the world."

Apart from issues such as those mentioned, which carry a strong message of enlightenment as well as a significant promise of being useful, there were many areas of scientific research that came to fruition in the 1960s in which useful applications turned out to be by far the most important byproduct. One thinks for example of the emergence of new synthetic polymers that have become an indispensable adjunct to industry and to modern living. Or the special roles that semiconductors, which were once regarded as chemical and physical curiosities, now play in our everyday lives. It was incidentally during this decade that several industrial research laboratories were struggling, ultimately with much success, to draw out the revolutionary promises inherent in the newly invented technology of integrated circuits. A key problem was to reduce production costs to the range required for everyday commercialization, in which event the influence of such circuits would presumably become "all pervasive" as Patrick Haggerty predicted and as ultimately proved to be the case.

Vice President Hubert H. Humphrey

Senator Hubert Humphrey became Vice President when President Johnson was elected to a second term in office in 1964. Humphrey possessed a lively and extensive interest in science and made multi-hour visits to the Academy on a fairly regular basis in order to catch up on anything new, potentially

Figure 41. Vice President Hubert H. Humphrey

exciting, or to raise questions about issues he had heard about by chance. It was a pleasure to be with him. He would have been a good friend of science had he been elected President in 1968. The State of Minnesota returned him to the Senate where he resumed many of his former activities.

The Enrico Fermi Laboratory

A crisis developed in the high-energy physics community in the mid 1960s. The Atomic Energy Commission had been using a committee of physicists under the chairmanship of Professor Norman Ramsey of Harvard University to make decisions concerning the type of high-energy accelerator that was most desirable to construct next. Seitz had served on the committee for a period and was familiar both with the individuals on it and with the issues under discussion. The most recent decision, supported by the AEC, had concluded that the next large accelerator should be constructed at the University of California at Berkeley, following a long succession of accelerators that had been created under the guidance of Ernest O. Lawrence, the inventor of the cyclotron. In the meantime, Lawrence had died, so new leadership was in charge.

The high-energy community was much disappointed when the proposed plans for the new accelerator were finally disclosed, since the new machine exhibited little imaginative flexibility and would clearly be primarily of local interest. Seitz began to receive visits from colleagues on the Ramsey committee who felt strongly that the decision to build the next device at Berkeley should be reopened and revalued. The scientific member of the AEC at that

Figure 42.   Glenn T. Seaborg

Figure 43. Emanuel Piore

time was Professor Glenn T. Seaborg, a distinguished Berkeley chemist. On visiting Seaborg, Seitz found that the commissioner was well aware of the complaints and had decided independently that they could not be ignored, even though contemplation of a review must have provided him painful hours. The new leaders at Berkeley were old friends and colleagues. Seaborg had shared a Nobel Prize with one of them.

In any event, the Academy created a new organization, The University Research Association (URA), with a small guiding staff and offered to use it as a base to review the situation. Dr. Emanuel R. Piore, a highly respected physicist who had played an important role in the evolution of the Office of Naval Research, was selected to chair the review committee, giving particular emphasis to the matter of the choice of site. Dr J. C. Warner, the president of what was then the Carnegie Institute of Technology, was selected to chair a committee that would determine the nature of the governing board of the organization for the accelerator. Early on, it had been agreed that the membership of the board would be national in scope and representative of the leadership in the high energy physics community from all parts of the country.

The Piore committee carried out a very thorough study of possible sites in cooperation with many experts. As that process reached its climax, Seitz received a call from President L. B. Johnson stating that when the list of acceptable sites had been reduced to the six best, he desired to make the final choice. He selected an acceptable site in Illinois, not far from Chicago and O'Hare Airport, which worked out very well.

At this point the Academy made URA an independent self-sustaining organization with the help of a member of its legal staff, Leonard Lee Bacon, who remained with URA. Its directors, in turn, selected Robert R. Wilson, then at Cornell University, as the director of the new laboratory.

Wilson and his colleagues provided excellent, highly imaginative leadership. The laboratory was eventually named after Enrico Fermi. It has had a very productive history. Unfortunately, later plans by URA to develop a larger successor accelerator in Texas did not receive continuing Congressional support. Leadership in the field of experimental high-energy physics will probably eventually go to the European laboratory centered in Geneva, Switzerland.

Aldabra Islands

A crisis of another kind arose as a result of a request for help from the Royal Society of London. Its leaders had learned that the Royal Air Force, using funding from United States, planned to establish an active airbase on the Aldabra Islands in the Indian Ocean in order to over-fly Africa for the purpose of gathering intelligence-oriented information. Up to that point the international scientific community had treated the relatively unpopulated islands informally as a very special nature reserve. As a result there was much concern about the inevitably significant ecological changes that would be produced by an airbase. The British government had rejected the objections advanced by the Royal Society, so its leaders turned to the U.S. National Academy for help. It would appear that for some reason or other the Royal Society had lost some of the influence it might have had on the government agencies at an earlier time. Seitz knew from his service on the Defense Science Board that the Pentagon was about to launch intelligence gathering satellites that would be much more effective. He discussed the matter with the Secretary of Defense, Robert McNamara, who turned the issue over to a young member of his staff, Rodney W. Nichols, with instructions to intervene in the funding of the British plan until the satellites were in operation. The islands were officially made a nature reserve in 1976.

Unidentified Flying Objects

Throughout human history some individuals have come to believe that they have experienced the supernatural or its equivalent. In our own time such experiences are often associated with the appearance of creatures from another planet either in our own or in another galaxy. They move about in machines of very advanced construction, popularly known as flying saucers, but technically designated as Unidentified Flying Objects, or UFOs. While many of the individuals who have had this experience or "sighting" are airplane pilots, they actually represent a small, although significant, component of the group that claims to have witnessed UFOs.

Figure 44.   Edward U. Condon

In 1965, the U.S. Air Force decided that it would try to get to the bottom of the matter by having a thorough study of the issue made by a distinguished group of scientists that would serve under the chairmanship of an individual who was widely respected within the scientific community. To be assured of retaining some degree of direct contact with the study, they would fund it through the institution, presumably a university, in which the chairman was located. Their representatives did, however, ask the Academy for advice in the selection of the chairman. While mulling the situation over with a number of colleagues, Seitz finally decided, with some reluctance, that the best person would be Professor Edward U. Condon, then on the faculty of the University of Colorado at Boulder, who was a deeply admired friend and counselor of long standing.

Early on, in his college days, Condon had been a colorful newspaper reporter before turning to physics as a profession. In the latter role he had earned world-class status. The reluctance to call on him stemmed from the fact that he had just passed through a very stormy decade during which he had been driven from the position as head of the National Bureau of Standards by a powerful group in Congress that resented his strong support of a movement favoring civilian rather than military control of the uses of nuclear energy, a movement that eventually led to the creation of the Atomic Energy Commission. He had finally found a refuge back in the relatively quiet academic world where he had begun and had every reason to avoid any activity that might lead to bitter controversy.

As it happened, Condon accepted the assignment with considerable enthusiasm, and helped in the selection of a superb group of scientists who formed the advisory study panel. The study extended over two years and led to a detailed report[7] that contained reviews of individual cases as well as general analysis. The chairman summarized the latter in eloquent and perceptive language. In many cases, the members of the advisory committee were able to provide probably correct explanations for the observations in what might be called everyday down-to-earth terms. It was, however, emphasized that the ability to use the scientific method in reaching a solution to a problem requires the ability to repeat in detail the circumstances under which any observation is made, a condition that is difficult to meet when dealing with individual random chance sightings, as is the case for UFOs. In other words, anything resembling a precise explanation for the phenomena related to the sighting of UFOs lies outside the framework of exact science at present. Nevertheless, the chairman and his committee urged the funding agencies to continue to support research in the field.

One should add that the two-year study was anything but tranquil. A very militant subgroup of those who were convinced that UFOs were of extraterrestrial origin also believed the government knew all the answers but was hiding the facts for its own presumably deceitful purpose. They regarded the study as a diversion and made a point of attacking Condon quite viciously. Fortunately, he stood up to the abuse although there were trying times. On observing the fracas, one news reporter said to Condon: "Doctor, in looking over your history, I note that you have been at the center of lots of trouble!" Condon replied: "It isn't hard."

Tricentennial of The French Academy Of Sciences

The French Academy of Sciences was founded in 1666, just after the Royal Society of London and the end of the religious wars. The French planned to have a significant celebration of the anniversary in 1966. Invitations were sent to science academies throughout the world welcoming representatives. Seitz accepted the invitation with appreciation of the fact that he would encounter one of the periods in which President Charles de Gaulle was venomously angry at the United States. Also it was likely that he would be the only U. S. citizen there who had actually been born and raised in United States, as turned out to be the case.

The opening ceremony was a highly colorful affair. President de Gaulle occupied an elevated throne-like chair, a military band played briskly in keeping with a prepared schedule, and the academicians entered the hall at the appointed moment wearing specially designed uniforms that were topped off by gallant cocked hats. Their marching pace did not conform even closely to the timing of the band. As a result, the band was replaced by a chamber orchestra on subsequent formal entrances of the academicians.

Seitz soon found that, with a few notable exceptions, even old friends conspicuously avoided him. As a result, he had much free time between meetings that were spread out over a week to visit museums, the palaces, and other old haunts in wonderful Paris. Thirty years later, he attended a medal-presenting meeting of the French Academy in conjunction with awards given to several young biochemists by a small American foundation. The mood and behavior of those involved in the gathering on that occasion was much different, resembling closely what one would have expected at a similar convivial meeting taking place at 21st and Constitution Avenue.

New Structures

In the meantime, there had been a renewal of the question of the way in which the professional engineers and the medical profession should be represented in the structure of the Academy. Both groups played major advisory roles and yet were in proportion greatly under-represented within the membership. The solution for engineers proved to be relatively simple, although it inevitably required much discussion. A leading group of engineers had decided to create an independent Academy of Engineering, but when offered the possibility by the Council of the National Academy of Sciences to join the existing National Academy under the umbrella of the original charter of 1863, it accepted, all with the understanding that in addition to electing its own professional membership it would have appropriate representation on the Governing Board of the National Research Council and on the Committee on Science and Public Policy. In the discussions, there had been some concern regarding the actions Congress might have taken if it had had freedom of choice in the matter. The proposal that the two academies unite was probably first made by the ongoing executive officer, S. Douglas Cornell.

The ranking members of the medical profession did not have a built-in solution comparable to that of the engineers to offer. Fortunately, Dr. James Shannon had just retired from the National Institutes of Health and was prepared to serve as special advisor to the Academy on this matter. He favored an organization somewhat parallel to that of the engineers in which the elected membership would be composed of distinguished medical doctors in various categories, including those involved in clinical practice and in research. It would also be represented on the governing board of the National Research Council. Shannon apparently foresaw the problems the medical profession would face in the years ahead as a result of matters such as the increased socialization of medicine and ethical issues associated with applications of some advances in biochemistry. He felt it would be good to have a strong organization representing the medical profession in Washington.

Figure 45.   James A. Shannon

James A. Shannon

Shannon (1904–1994) had a dynamic career. He started medical research in the field of renal physiology, but shifted to the search for anti-malarial drugs during World War II. At war's end, he accepted a position as head of an industrial pharmaceutical laboratory for four years, gaining special experience that served him well in his later public services. In 1949, he joined the Institutes of Health (NIH), a relatively modest organization at the time, reporting to the Surgeon General in the Public Health Service. There he became a member of the National Heart Institute and was placed in charge of its research program in 1953 when NIH was absorbed into the newly created Department of Health Education and Welfare. His remarkable qualities of leadership soon became generally evident and he was selected to be the director of The National Institutes of Health when the position became open in 1955. With the skillfully won cooperation of some powerful members of the Senate and House, as well as the Executive Office, he started the transformation of the NIH into the relatively well-funded, essentially all-pervasive institution we know today. In setting very high standards of professional achievement for NIH, Shannon, somewhat like John Powell, made enemies of some individuals on the Hill who had more mundane goals of their own. As in Powell's case, they were able to force him from office somewhat prematurely once his more powerful supporters departed from the Hill.

Shannon left Washington for a position in New York in 1969 and turned the matter of developing the relationships between the medical community and the Academy over to an old medical friend, Dr. Walsh McDermott, who greatly desired to be involved in laying the foundations for the new addition to the Academy. McDermott, along with colleagues, developed a much more diverse and cosmopolitan structure for the new entity, now known as the Institute of Medicine (IOM), than that envisioned by Shannon. It was attached to the Academy. The acceptance of the McDermott plan by the Council occurred during the presidency of Professor Philip Handler (1969–1981). Shannon was not at all pleased with the result at the time and threatened not to accept the status of a charter member. For one reason or another he did not follow through on this threat and is listed among the founding members. The first president of the IOM was Professor John Hogness, who served between 1972 and 1974. (See Appendix B). McDermott (1909–1981) had a distinguished research career in the Cornell Medical School working on the development of antibiotics, some of which was carried out in cooperation with René Dubos of The Rockefeller Institute across the street. He had an essentially life-long struggle with tuberculosis and benefited from his own research.

Office of Technology Assessment

For many decades, both the scientific community and the members of Congress involved in scientific affairs had wondered if it would be possible to create a committee or board that might be able to forecast the effects that some newly introduced technical development could eventually have upon society. The concept is not unreasonable as is demonstrated by the fact that individuals such as Patrick E. Haggerty and Gordon E. Moore were able to make fairly good estimates of the ultimate consequences of the introduction of semiconductor-based electronics quite soon after the invention of the integrated circuit. Finally, two members of the House took the lead in the matter by preparing and introducing a bill that would create an Office of Technology Assessment (OTA) linked to the Hill. The two had different party affiliations, but they realized full well that good science is non-partisan. One of the individuals was Emilio Daddario (Democrat) and the other Charles Mosher (Republican). A companion bill was introduced in the Senate.

Their planning began early in the 1960s and continued throughout the decade. At the time, Representative Daddario was chairman of the House Subcommittee on Science Research and Development and had a remarkably clear concept of the place of science in the underpinnings of our society. The pair recognized among other things that the ultimate effects of the applications of science could have deleterious side effects as well as great benefits and hoped that the former could be minimized if revealed sufficiently early. Seitz spent much time with them discussing the details of their bill in the course of its creation.

At the core of the original plan was a Technical Assessment Board (TAB) consisting of six senators and six representatives, equally divided between both political parties in recognition of the fact that basic science should be apolitical. The chairmanship and vice chairmanship of the TAB would alternate between Senate and House in succeeding Congresses. The TAB would appoint a director of the OTA to serve for a six year term and would also appoint an Advisory Council (AC) of ten eminent citizens from industry, academia or elsewhere outside the federal government to advise the agency. In the original bill, four of these were to be appointed by the president of the United States. Advice could also be sought from outside the AC. The comptroller general of the United States and the Congressional Research Service served as statutory members. The OTA would take on assessments at the request of the chairman of any congressional committee. The chairman could request the work personally, on behalf of a ranking minority member or on behalf of a majority of the members of a committee. The Technology Assessment Board could request work, as could the director. The OTA would acquire a staff as determined by budgets and requests.

Compatible bills for House and Senate were ready by 1970, but were held over for voting in 1972. The House passed the bill essentially in the form developed by Daddario and Mosher, but the Senate insisted on significant changes at the request of Senator Jack Brooks from Texas, in consultation with colleagues in the Senate. His amendment required the number of members of the TAB from each house be reduced to five, three from the majority party and two from the minority. Also it required that the revised TAB select the director. These changes placed most of the authority for controlling the OTA in the hands of the two houses and in the hands of the majority party when it controlled both. One might have thought that the Republican leadership would have wanted to retain the Office when it regained control of the two houses, but apparently the party leadership felt otherwise in 1995.

In the meantime Daddario was no longer in the House of Representatives and did not play a formal role in the final decisions relating to the creation of the OTA. At the request of his party, he had run for the position of governor of his home state of Connecticut in 1970 and failed to win. He was appointed director of the Office of Technology Assessment in 1973, holding the post until 1977, and was president of the AAAS for a two-year term (1977–1978).

A New Auditorium

What remained to be accomplished in a major way in the improvement of facilities at 21st and Constitution Avenue was the addition of a suitable auditorium.[8] For several decades after 1922 a small lecture room adjacent to the

Board Room had served most needs for convocations of the academy members, including the gatherings at which new members were elected. This proved inadequate after World War II as the membership grew rapidly in size. The Great Hall soon began to provide a substitute. It, however, did not have good acoustical properties and required restaging for each important meeting in which it served as a lecture hall. There was adequate room for such an addition in the space between the East and West Wings that had been added by Bronk. Unfortunately, a sum between four and five million dollars was required to achieve the goal. A very special plan would be needed if the campaign seeking the funds was to be successful.

Hugh L. Dryden

Dr. Hugh Dryden (1898–1965), a native of Maryland, was one of the leading aeronautical engineers in the country. He had completed his formal advanced education in his professional field at The Johns Hopkins University in 1922 and joined what was then the National Bureau of Standards as head of the aerodynamics section. He carried out advanced research in the sonic and turbulent ranges of airflow, helping to pave the way for what became supersonic flight. During World War II, he worked with the NDRC on the use of armed live bats as guided missiles in defense against hostile Japanese aircraft, a program that was sufficiently successful to gain commendation. This was one of the several methods developed for such defense, the most successful being the proximity fuse developed by Merle Tuve with the cooperation of the Navy under the guidance of Captain William S. Parsons and based on the use of a

Figure 46.　Hugh L. Dryden

miniature microwave radar unit imbedded in the weapon. It was usually fired from a gun. Parsons incidentally was later the officer in charge of the arming in flight of the nuclear bomb dropped on Hiroshima.

In 1947 Dryden became director of flight of the National Advisory Committee on Aeronautics (NACA), which had been created in World War I and was central for much of planning in the field of aviation in the country over the years. He was appointed director of the entire organization two years later.

When as mentioned earlier the Soviet Union launched the small earth-orbiting satellite *Sputnik* in the autumn of 1957, the U.S. government terminated NACA and, in 1958, passed the legislation needed for its replacement, the National Aeronautics and Space Administration (NASA). Dryden drafted the necessary documents, which were accepted in the form he presented with relatively little change. Soon after, he began encouraging studies of the nature of the heat shields that would be required on some rockets during re-entry into the atmosphere from very high orbit.

Although the White House and Congress recognized the sterling scientific, technical and organizational capabilities that Dryden possessed, they were not prepared to select him as administrator of NASA. His natural approaches to matters associated with public relations were reserved and non-partisan, being conditioned both by his basic personality, which was low-key, and many years as a reserved public servant. In consequence the position of administrator of the new agency was left in temporary hands, namely those of Dr. T. Keith Glennan, the president of what was now the Case Western Reserve University, for a year or so while awaiting a newly elected president of the United States. Dr. Glennan immediately offered the position of associate administrator to Dryden. In the meantime Dryden had also been offered several major non-governmental positions, such as a professorship at the Massachusetts Institute of Technology, but he had become so devoted to the promises NASA offered for new activities in space that he decided to remain in Washington.

Dryden had been elected Home Secretary of the Academy in 1955 and felt sufficiently honored by the post that he maintained it until his death ten years later. He revered the academy and his association with it so much that he was happy to carry on with the additional burden in spite of his heavy duties at NASA.

The new president of the United States, John F. Kennedy, decided that one of the primary missions of NASA should be to land astronauts on the moon, the challenge for a great adventure.

James E. Webb

James E. Webb (1906–1992), mentioned earlier, was one of the most remarkable individuals of the many who seek a career in Washington. Mervin

Figure 47. James E. Webb

Kelly, the president of the Bell Telephone Laboratories, once commented that anyone who has been associated with Webb in a significant way would have their life changed irreversibly, a situation to which many can attest. It was also a statement that could be applied to Kelly. In keeping with his ambition for a political career, Webb studied law, passed the Washington bar examinations in 1936 and began to enter practice. It was interrupted by a period in the Marine Corps but at the end of the war, in 1946, he became undersecretary of the Treasury, which was followed by a period as head of the Bureau of the Budget.

Webb left Washington during the Eisenhower administration to take a sequence of private positions, mainly with industry, but returned in 1960 to seek a post in the Kennedy administration. By this time, his many special aptitudes were widely appreciated and President Kennedy offered him the now vacant position of administrator of NASA. He accepted under the condition that the program be regarded as the start of a continuing one, not focused only on a single successful voyage for short-term propaganda purposes. President Kennedy agreed.

As mentioned earlier, the Academy had formed an advisory board on space science under Lloyd Berkner soon after the appearance of *Sputnik*. It had an enthusiastic constituency that was greatly stimulated by discoveries such as X-ray emitting stars and the ionization belts around the earth (the Van Allen Belts), which were being made with instrumented space probes. The committee soon began to receive financial and other support from NASA. President Kennedy's decision to send astronauts to the moon, however, caused a split within the constituency since a substantial fraction of the group would have preferred to see NASA focus, at least initially, on unmanned missions

that possess high scientific value, such as continuing the study of X-ray-pro-
ducing stars. Another component of the committee was sympathetic to President Kennedy's proposal since it was prepared to recognize the values the
plan possessed for international publicity and domestic politics, if successful.
They also realized that success in a manned program could stimulate public
interest in associated scientific research.

It so happened that the chairman of the Academy's space committee who
followed Berkner strongly opposed the president's plan. Moreover, he was
quite outspoken about expressing his opinions. Webb quickly learned of this
and asked Seitz if the chairman of the committee, a brilliantly productive scientist, could be replaced. The response had to be: "Jim, the membership
would replace me first!"

Webb had the gift of judging the merits of individuals in other professions
rapidly and fully appreciated the talents of Dryden, whom he asked to continue on as deputy administrator. The two were perfect complements in their
posts, and much of the success of the Apollo program can be ascribed to their
close partnership. It lasted until the end of 1965 when Dryden finally suc-
cumbed to a slowly developing cancer for which he had been receiving steady
treatment while continuing to work industriously. In the meantime, the pattern of operation necessary for the success of the Apollo program had been
well established. Since Webb had become very close to Dryden and missed
his presence keenly on both the personal and technical sides, he felt that some
special steps should be taken to honor his former colleague.

As a result of several conversations between Webb and Seitz, a decision
was made to raise a fund in Dryden's name to construct the auditorium. Once
the basic idea had taken hold, the council of the Academy appointed a small
guiding committee composed of individuals who had known and admired
Dryden both as a personal friend and highly experienced engineer and ad-
ministrator. The initial group consisted of Donald W. Douglas, Jr. (chairman),
Allen V. Astin, Detlev W. Bronk, Leonard Carmichael, James H. Doolittle,
Harry F. Guggenheim, S. Paul Johnston, Augustus B. Kinzel, Vice Admiral
Emory S. Land, U.S.N. (ret.), Grover Loening, Merle A. Tuve, and James E.
Webb.

John S. Coleman, the executive officer of the Academy at the time, became
as deeply involved in the program as any other participant and was to no
small degree responsible for the success achieved. Additional names were
added to the advisory committee as the raising of funds progressed and new
opportunities arose.

It was hoped that a substantial amount of the needed money could be derived from the aerospace industry, which had benefited so much from Dryden's lifetime involvement in the field as well as his work for NASA. Indus-

try ended up providing about twenty percent of the funds needed. The greatest fraction came from the private foundations, about one third. The membership of the Academy was also very generous within its means. Webb contributed all of the lecture fees he received from the many public addresses that he gave concerning the development of the Apollo program.

As the final goal of the fund drive began to be reached, the choice of new sources began to shrink and the Academy began to draw on some of its relatively small reserve funds. Fortunately, the National Institutes of Health (NIH) came to the rescue at a critical moment, just as the hazard of serious inflation began to loom on the horizon, along with an associated increase in construction costs. It appeared that one might be pursuing the end of a rainbow. Seitz visited Dr. James Shannon and discussed the shortfall. Shannon asked if it would be possible to designate a special area in the structure that would be created for use in the extensive medical advisory work of the Academy. A discussion with the architect demonstrated that this was quite feasible without radical changes in plans, whereupon Shannon gathered together his staff, popularly known as "the Irish Mafia," and with its help in the review of budgets was able to contribute seven hundred and fifty thousand dollars, which made it possible to sign a contract for construction.

Once the critical sum of money needed to provide a useful auditorium had been reached, Dr. William O. Baker, a member of the academy, president of the Bell Telephone Laboratories and personal advisor to Mr. Paul Mellon on special philanthropic matters, advised the latter to add five hundred thousand dollars to the fund "to ensure that the architectural style of the addition matches the high level of the existing structures in the complex."

The selection of an architect placed Seitz in an awkward position. The logical choice was Wallace Harrison, who had played such a significant role as a young man in the design of the original building under the guidance of Clarence Goodhue and later in the design of the East and West Wings. Bronk had also employed Harrison in the upgrading of the campus of The Rockefeller University once he became president there. However, Bronk, a man occasionally short of temper, had a serious dispute with the architect over the not entirely surprising failure of a novel form of ceramic tile roofing on the University's newly-constructed auditorium. A long-standing productive friendship was seriously disrupted. Bronk felt that Harrison might well prove to lack interest in planning the auditorium. Yet Seitz decided that it would be improper to ignore Harrison and arranged a meeting to discuss the issue. Harrison virtually jumped at the opportunity, partly for sentimental reasons and partly to renew close associations with his architectural friends and colleagues in Washington who would review his plans for approval.

The program of design moved rapidly and the groundbreaking ceremony for the auditorium took place on August 1, 1968. The construction began immediately after and was completed at the end of 1970. The dedication took place at the annual meeting in the following April, during the presidency of Philip Handler.

James Webb expressed disappointment that Hugh Dryden's name was not more prominently displayed in connection with the auditorium. In reviewing the matter, the council of the Academy feared at the time that if segments of the Academy building began to receive special names associated with donors, the edifice might soon become cluttered with name plaques and lose much of the distinctiveness associated with its primary devotion to science. As a result, the Academy adopted a rule, valid for the period, that no nameplates would be added to the building. There was to be only one exception, namely the West Wing, which had an understandable origin. Dryden, above all, would have understood.

In the meantime, the council of the Academy has had second thoughts about the matter of increasing its private endowment by various means, including affixing the names of appropriate prominent donors to portions of the headquarters complex on Constitution Avenue. Under the proper circumstances, this could provide added freedom from politically motivated influence originating from the government and alleviate, at least in part, Joseph Henry's great concern about the ultimate effectiveness of the institution as a result of its close ties to government.

PHILIP HANDLER

Philip Handler (1917–1981), a native New Yorker, was not only a dedicated scientist, but displayed an unusually high degree of sensitivity in his responses to situations involving conflicting opinions. He was ideally suited to lead the Academy through the tempestuous political climate associated with the war in Vietnam to which many members of the academic community were strongly opposed. He succeeded in bringing an essential degree of unity to the membership by emphasizing the great need for universal support of the principle of human rights. He delivered many eloquent addresses on this topic from the podium of the new auditorium as well as at many other places at home and abroad. To express his own general range of sympathy with the spirit of those who opposed the war in Vietnam, he did not exercise his statutory seat on the Defense Science Board.

This is not to say that Handler's period in office was an easy one. A small but highly activist group of members of the Academy harried him incessantly

Figure 48.   Philip Handler. Photo by Antony Dr. Gesu.

in an attempt to have the Academy express open opposition to the actions of the government but he was not prepared to do this. Fortunately, the majority of the membership supported him.

When Handler retired in 1981, there was a move on the part of his justly admiring friends to name the auditorium after him. President Frank Press, who followed Handler, emphasized that the funds that were raised to construct the auditorium had been carried out in the name of Hugh Dryden so that if any name were used, it would necessarily be Dryden's. A symbol of Dryden's role is provided by a bronze copy of his bust that is situated in a special display room at the rear upper level of the auditorium.

Since his family had limited means during the Depression, Handler entered City College in New York in 1933 and graduated three years later. His age was out of phase with that of most of his classmates so he found time to engage in self analysis and speculate on plans for the future. One of the members of the faculty of the college who noted that Handler was excellent in chemistry recommended that he enter graduate school at the University of Illinois, which had a very strong department in organic chemistry. Arrangements were made whereby he received a paying half-time position as a chemical assistant in the School of Agriculture and was allowed to register as a graduate student in the department of chemistry. He soon attached himself to a new young member of the faculty, Herbert E. Carter, who was engaged in the field of biochemistry, which was beginning to attract much attention. He and Carter, who quickly came to admire Handler's many gifts, worked well together and became close, lifetime friends. The peripheral experience he gained working in agricultural chemistry added an important practical twist to Handler's education.

Handler completed graduate work in 1939. Since any thought of a fellowship abroad seemed impractical as war clouds gathered, he accepted a position at Duke University, which proved to be his base for forty-two years. He developed deep roots there, but also became a prominent figure on both the national and international scenes. He was particularly admired in Washington and was frequently asked to testify before congressional committees on scientific matters related to his areas of professional involvement. He was the very eloquent "Dr. Science" to many on the Hill long before he became president of the Academy. James Shannon, the head of NIH, must have appreciated his fine support.

Handler was elected president in 1968 but had to delay taking office for a year for personal reasons, during which Seitz remained in the chair. In his twelve years of service, Handler continued the evolution of the Academy's activities on the national and international frontiers of science that had started in the Bronk era. He received many honors and became a celebrated figure in the larger Washington community. He accepted membership on the board of what had become The Rockefeller University.

Copernicus Celebration

The semi-millennial anniversary of Copernicus' birth (1473) occurred during Handler's period in office. He made certain that the event was celebrated appropriately during a yearlong American festival.

He also commissioned a life-sized statue of Albert Einstein by a sculptor devoted at the time to a special style of contemporary art. It displays the immortal genius in his very late years seated before a small synthetic pool, presumably pondering the mysteries of our universe. The representation of him is similar to those commonly seen in many advertisements in newspapers, magazines or billboards in United States promoting some commercial device or plan. It probably is not one that would have been selected by a group of physicists familiar with his appearance (Figures 10 and 19) during his most creative three decades that began in his mid-twenties. The statue is however very popular with the general public in United States since the representation is one of a type they see frequently and is fixed in their minds. It is situated on the front lawn close to Constitution Avenue and serves as a temporary pausing point for the innumerable tour buses that circulate about Washington in season.

The Oklo "Natural" Nuclear Reactor

In 1972, soon after Handler took office, a group of French mining engineers made a most interesting discovery while exploiting a uranium mine on the

Oklo River in Gabon, Africa. They concluded that the uranium deposit had been the site of a spontaneous nuclear chain reaction some 1.8 billion years ago when the relative concentration of uranium 235 isotope was about four times greater that at present. Dr. Alvin M. Weinberg, the former director of Oak Ridge National laboratory has made a special analysis of the implications of this discovery, *Nature* 266(1977): 206.

Toward the end of his terms in office, Handler contracted lymphoma and apparently tried some form of self-treatment. By the time his medical colleagues appreciated the situation and insisted that he be placed in the hands of cancer specialists, it was too late.

FRANK PRESS

Philip Handler's successor, Frank Press (1924–), a geophysicist, was also a New Yorker. He was elected president of the Academy in 1981 and served for a full twelve years. He had also received his undergraduate education at what was then the City College of New York and had completed graduate work in geophysics at Columbia University in 1949. He remained at Columbia University in junior faculty positions until 1955 when he received a professorship at the California Institute of Technology, along with the directorship of the Seismology Laboratory of the Institute. In 1965 he was appointed professor of geophysics and chairman of the department of earth and planetary sciences at the Massachusetts Institute of Technology. This appointment was interrupted by a call to Washington, D. C., in 1977. There he served an

Figure 49.   Frank Press

eventful four year term as science advisor to President Jimmy Carter and director of the Office of Science and Technology Policy in the Executive Offices of the White House. He was elected president of the Academy soon after returning to Cambridge at the end of the Carter administration with the result that he reversed his steps, back to Washington.

Frank Press' career is a highly distinguished one on many counts. At the scientific base, he belonged to the first generation of geophysicists who benefited from and contributed to the magnificent revolutionary developments associated with the evolution of the field of plate tectonics, briefly described earlier in this document. He became one of the leaders in this area of research and received appropriate recognition both domestically and abroad. For example, he was awarded the gold medal of the Royal Astronomical Society in 1972 and the U. S. National Medal of Science in 1994. Similar recognition for his scientific achievements was provided by France, Germany, Japan and Russia over the years. More specifically, he was one of the two individuals awarded the Japan Prize in 1993. The prize is regarded to be at the same level as the Nobel Prize in the fields it recognizes and is presented in person by the Emperor.

On retiring at the age of 69 in 1993 from a very active period of leadership in the Academy, he decided not to return to an academic position at a university, as might have been feasible, but to accept a four year appointment (1993–1997) as the Cecil and Ida Green Research fellow at the Carnegie Institution in Washington. This left him with much freedom for action. He did, however, establish formal links that were partly, but by no means entirely, honorific, with Columbia University, the California Institute of Technology and the Massachusetts Institute of Technology.

In his new role, Press soon found himself called upon as a wise and highly informed advisor by many organizations, both public and private. The magnitude and diversity of the overall requests turned out to be sufficiently great that it proved desirable not only to join with others in formulating responses, but to create at least one flexible institution with a formal business structure.

The innumerable effective forms of leadership that Press has displayed over the years from his bases in Washington, and in his quiet and reserved manner, stand up well when compared with those of Joseph Henry and Charles Walcott. In the year 2000 he was designated "President Emeritus of the National Academy of Sciences", apparently the first time the title was granted. It is however one that Detlev Bronk would have appreciated possessing in his later years. The latter retained a working office at the Academy as well as at The Rockefeller University until his death.

BRUCE MICHAEL ALBERTS

Frank Press was followed in 1993 by a biochemist, Bruce M. Alberts (1938–) who was born in Chicago, Illinois. The latter entered Harvard as an undergraduate in the field of biochemical sciences in 1956 just as molecular biology was beginning to blossom as a result of the unraveling of the genetic code and the application of the knowledge gained thereby. He continued on for graduate work at Harvard, obtaining a doctor's degree in biophysics in 1965. He was then granted a National Science Foundation postdoctoral fellowship that permitted him to spend a year at the Institut de Biologie Moleculaire in Geneva, Switzerland. The fellowship was followed by a stepwise series of posts at Princeton University which led to the Damon Pfeiffer Professorship in life sciences. He held the latter between 1973 and 1976.

In 1976 he left Princeton for the University of California in San Francisco with an initial appointment as professor and vice chairman of the department of biochemistry and biophysics. In 1981 he was appointed to an American Cancer Society Research Professorship, which he held actively until 1993 when not serving as department chairman. The title was granted for life in 1980.

Well before leaving for Washington in 1993 Alberts was recognized universally as one of the very capable and imaginative leaders in the field of molecular biology. As a result he began to receive appointments to public and private boards and committees that are active in the advance of the profession. Such service, which grew almost without limit once he became

Figure 50.   Bruce M. Alberts

president of the Academy, has occupied an important fraction of his career and is responsible for some of the distinguished awards he has received.

Alberts' twelve years in office are notable for the more or less final resolution of several major issues related to the operation of the Academy that had been dangling somewhat indecisively for several generations. Three of the most significant are as follows.

Endowment

The first and in a sense the most radical is an enlarged emphasis on raising sufficient private endowment from whatever sources are available to permit the organization to provide advice on any topic that lies within its sphere of interest without the need to call upon federal funds. The completion of a five-year joint campaign for the National Academies that raised more than 300 million dollars from private sources is, in effect, at least a partial declaration of independence from the federal government, with residual ties as expressed in its charter of 1863 to serve that government on request. Circumstances indicate that the Academy can indeed raise a substantial endowment. Guidance provided by wise leadership will be required to determine the manner in which the enhanced authority is employed. There is the hazard that the academy could eventually become looked upon as another one of the almost unlimited number of Washington think tanks having its own agenda.

Foreign Academies

It will be recalled that when Detlev Bronk appointed Harrison Brown foreign secretary of the Academy he gave the latter the authority to expand the functions of the office by forming associations with national or regional academies well outside Europe. Brown took full advantage of the enlarged responsibility by increasing the size of the staff and the number of experienced advisors. He also formed teams or "desks" which developed continuing links with groups of academies that were of special interest to his advisors. The main obstacles to this venturesome activity, apart from limitations of money, arose from the fact that the ongoing Cold War complicated communication in such a way that Brown and his staff did not always have complete freedom of action to the degree he would have preferred. Finally, ill health impeded his personal involvement.

Fortunately Brown's good work and persistence did make it possible to create, among others, two standing committees which have had an enduring influence and fit well into more recent developments, namely The Standing Committee on the People's Republic of China and The Standing Committee on Less Developed Countries.

The end of the Cold War, coupled with the widespread increase in understanding of the importance of science in the political, social, medical and economic aspects of life made it possible for Alberts and his colleagues to form the Interacademy Panel (IAP) of more than ninety science academies with its secretariat initially at the Royal Society in London, and since 2000 at the Third World Academy of Sciences (WAS) in Trieste, Italy. The IAP holds promise of developing a very practical operating network wherein the stronger institutions provide genuine help to the weaker while broadening their own horizons.

The IAP in turn formed the Interacademy Council (IAC) in 2000 with its secretariat at the Royal Netherlands Academy of Arts and Sciences in Amsterdam. The AIC is governed by the presidents of fifteen science academies. Alberts serves as one of its co-chairs for the period from 2000 to 2009. Modeled after the National Research Council, the IAC serves to mobilize the best science for world decision makers. Its first report entitled "Inventing a Better Future: A Strategy for Building World-Wide Capacities in Science and Technology" was released by Secretary General Kofi Annan at the United Nations in 2004.

Housing The National Research Council

Although the executive offices of the National Academy of Sciences occupy fine quarters in the main building on Constitution Avenue, housing for the activities of the National Research Council has offered a continuing challenge. Initially much of the work that could not be squeezed into the main building somehow was distributed throughout Washington in rented offices and the like as circumstances permitted. As mentioned earlier, some of the pressure was relieved, first by the construction of the East and West Wings of the main building and then by the leasing of a specially constructed building on Pennsylvania Avenue from George Washington University in the 1960s. When the lease ran out during the terms of Frank Press, a building complex on Wisconsin Avenue bearing special financial benefits was found. It served adequately for a number of years, but then proved insufficient for the ever-growing needs during Alberts' period in office. On this occasion it proved possible to obtain a relatively elegant solution, namely to design and build the Keck building in downtown Washington through a combination of Academy resources, philanthropy and careful planning. With the capacity to house 1000 staff, one might hope that it will serve the needs for most of the present century.

An Emphasis on Education

The National Research Council successfully completed the four–year process of producing the first-ever National Education Standards for the United States in 1996. The massive effort ushered in a major expansion of activities in education, with a focus on using analysis and evidence to create a continuously improving education system from kindergarten through graduate school. A new Center for Education was established, with separate boards for science, mathematics, and testing and assessment. Although the focus was on improving education in science and mathematics, highly successful studies were also carried out on reading, on how people learn and strategies for effective research in education.

Bruce Alberts The Teacher

Bruce is at heart a great teacher of science with a very special approach to that art. He admits that his greatest regret on departing for Washington from San Francisco in 1993 was to leave behind the continuous association with students, both young and old, that one encounters in a lively university setting. In compensation he had the hope that his terms in Washington would make it possible to have a significant influence on the national standards of education, not least those for science. And he worked unceasingly toward that goal both openly and behind the scenes.

His personal approach to education in science is not to ply the student with a tidal wave of rote, but rather to challenge his or her mind with a paradox or puzzle drawn from the heart of good science that will offer a catalytic start in imaginative thinking of the kind that breeds new science.

After completing twelve years as president of the Academy, he has returned to the University of California in San Francisco as a professor in the department of biochemistry and biophysics. The students in San Francisco will be fortunate when he again picks up the traces there.

RALPH JOHN CICERONE

Ralph Cicerone (1943-) was elected president of the Academy for an initial, renewable, term of six years in 2005. He had been elected to the Academy in 1990. As president, he brought with him a very rich and diverse background of experience, although his principal scientific interests had become focused on atmospheric chemistry, geophysics, and environmental science by 2005.

He was born in Pennsylvania and apparently decided initially on a physical and technical career since his choice for undergraduate study was the Massachusetts Institute of Technology, which he completed in 1965. He then proceeded with graduate work at the University of Illinois, obtaining a Ph.D.

Figure 51.   Ralph Cicerone

degree in electrical engineering with a minor in physics in 1970. He obtained his first exposure to aeronomy during this period. His first post-doctoral academic appointments were at the University of Michigan at Ann Arbor and extended over the period from 1970 to 1978. Here he expanded the scientific background at his command to chemistry, particularly atmospheric chemistry. By 1978 he was committed to environmental studies with central emphasis on the chemistry of the ozone layer and factors that had caused a period of global warming starting in the nineteenth century. The years 1978 to 1989 were spent in close association with the Scripps Institution of Oceanography in La Jolla and the National Center for Atmospheric Research in Boulder, Colorado, where he carried out much of the research upon which his scientific reputation is based, including the sources of atmospheric methane and halogen-containing chemicals in the atmosphere.

In 1989, he was appointed to the Daniel G. Aldrich, Jr. chair in earth system science at the University of California at Irvine and to a professorship in the department of chemistry. He became dean of physical sciences of the university in 1998. While demonstrating appropriate leadership in his administrative posts at the university, Cicerone continued to maintain an active research laboratory and serve as international spokesman on environmental science. He also chaired several studies of environmental issues at the National Academy of Sciences. For these activities he was granted a number of prestigious awards such as the Revelle Medal of the American Geophysical Union, the Bower Award of the Franklin Institute and the Einstein Medal for Science from the World Cultural Council.

At the annual spring meeting of the Academy in April of 2006, a year after his election, President Cicerone delivered his first annual review of on-going activities and the ever-growing challenges to be faced in the future. The guiding theme of his address stressed the fact that the mission of the institution is both "timeless and timely." The address made it evident that he was now as fully in control of his complex office as a president can hope to be.

Appendix A

Source of Photographs

1) Alexander D. Bache. Archives of The National Academy of Sciences (NAS).
2) Joseph Henry. Archives of NAS.
3) William B. Rogers. Archives of NAS.
4) John Wesley Powell. Archives of NAS.
5) Othniel C. Marsh. Archives of NAS.
6) Henry A. Rowland. Courtesy of The Johns Hopkins University Archives and the Emilio Segré Visual Archives of the American Institute of Physics.
7) Wolcott Gibbs. Archives of NAS.
8) Alexander Agassiz. Archives of NAS.
9) Louis Agassiz. Courtesy of the Museum of Comparative Zoology, Harvard University.
10) Albert Einstein. Getty Archives, courtesy of Dr. Qinghong Yang.
11) Ira Remsen. Archives of NAS.
12) George E. Hale. Archives of NAS.
13) William D. Harkins. Archives of NAS.
14) William H. Welch. Archives of NAS.
15) Simon Flexner. The Rockefeller Archives, Pocantico, New York.
16) Hideyo Noguchi. Courtesy of Mrs. Qinhong Yang.
17) Charles D. Wolcott. Courtesy of the Smithsonian Institution.
18) Robert A. Millikan. Archives of NAS.
19) Albert Einstein. Courtesy of American Institute of Physics Emilio Segré Visual Archives, W. F. Meggers Gallery of Nobel Laureates.
20) Albert A. Michelson. Courtesy of the Michelson Museum, Naval Weapons Center, China Lake, California, and NAS Archives.

21) Thomas H. Morgan. Archives of NAS.
22) William W. Campbell. Sketch by Peter Van Valkenburgh. Archives of NAS.
23) James Lick. Courtesy of Lick-Wilmerding High School in San Francisco and Headmaster Dr. Albert Adams.
24) George Davidson. Archives of NAS.
25) Benjamin A. Gould. Archives of NAS.
26) Frank R. Lillie. Archives of NAS.
27) Frank B. Jewett. Archives of NAS.
28) Vannevar Bush. Archives of NAS.
29) James B. Conant. Courtesy of Harvard University. Photo by Karsh of Ottawa.
30) Detlev W. Bronk. The Rockefeller University Archives.
31) Alfred N. Richards. Archives of NAS.
32) Mervin J. Kelly. Courtesy of the Emilio Segré Visual Archives of the American Institute of Physics. Photo by Werner Wolff.
33) Patrick E. Haggerty, with David Rockefeller. The Rockefeller University Archives.
34) Harrison Brown. Archives of NAS.
35) Lloyd V. Berkner. Courtesy of the Emilio Segré Visual Archives of the American Institute of Physics.
36) George B. Kistiakowsky. Courtesy of the Emilio Segré Visual Archives of the American Institute of Physics.
37) Frederick Seitz. The Rockefeller University Archives.
38) Maclyn McCarty. The Rockefeller University Archives. Photo by Ingbert Gruettner.
39) Harry H. Hess. Courtesy of Princeton University Library. University Archives. Department of Rare Books and Special Collections.
40) Roger Revelle. Courtesy of the Emilio Segré Visual Archives of the American Institute of Physics.
41) Hubert H. Humphrey. Courtesy of his son, H. H. Humphrey III.
42) Glenn T. Seaborg. Courtesy of Atomic Energy Commission and NAS Archives.
43) Emanuel Piore. Courtesy of the Emilio Segré Visual Archives of the American Institute of Physics.
44) Edward U. Condon. American Institute of Physics Gallery of Member Society Presidents.
45) James A. Shannon. Archives of NAS. Photo by Edward A. Hubbard.
46) Hugh L. Dryden. Courtesy of the National Aeronautics and Space Administration and the Emilio Segré Visual Archives of the American Institute of Physics.

47) James E. Webb. Courtesy of the National Aeronautics and Space Administration.
48) Philip Handler. Archives of NAS.
49) Frank Press. Archives of NAS.
50) Bruce Michael Alberts. Archives of NAS.
51) Ralph John Cicerone. Archives of NAS.
52) John Wesley Powell in conversation with a Paiute native American. Courtesy of the Smithsonian Institution and Dr. Marc Rothenberg.

Appendix B

Officers of the National Academy

VICE PRESIDENTS

1863–1865	James Dwight Dana
1866–1868	Joseph Henry
1868–1871	William Chauvenet
1872–1878	Wolcott Gibbs
1878–1883	Othniel Charles Marsh
1883–1889	Simon Newcomb
1889–1891	Samuel Pierpont Langley
1891–1897	Francis Amasa Walker
1897–1903	Asaph Hall
1903–1907	Ira Remsen
1907–1917	Charles Doolittle Walcott
1917–1923	Albert Abraham Michelson
1923 1927	John Campbell Merriam
1927–1931	Frederick Eugene Wright
1931–1933	David White
1933–1941	Arthur Louis Day
1941–1945	Isaiah Bowen
1945–1949	Luther Pfahler Eisenhart
1949–1953	Edwin Bidwell Wilson
1953–1957	George Washington Corner
1957–1961	Farrington Daniels
1961–1965	Julius Adams Stratton
1965–1973	George B. Kistiakowsky
1973–1981	Saunders MacLane

1981–1993   James David Ebert
1993–2001   Jack Halpern
2001–2005   James S. Langer
2005–   Barbara A. Schaal

FOREIGN SECRETARIES

1863–1873   Louis Agassiz
1874–1880   F. A. P. Barnard
1880–1886   Alexander Agassiz
1886–1895   Wolcott Gibbs
1895–1901   Alexander Agassiz
1901–1903   Ira Remsen
1903–1909   Simon Newcomb
1909–1910   Alexander Agassiz
1910–1921   George Ellery Hale
1921–1934   Robert Andrews Millikan
1934–1936   Thomas Hunt Morgan
1936–1942   Lawrence Joseph Henderson
1942–1945   Walter Bradford Cannon
1945–1950   Detlev Wulf Bronk
1950–1954   Roger Adams
1954–1958   John G. Kirkwood
1958–1961   H. P. Robertson
1962–1974   Harrison Scott Brown
1974–1978   George S. Hammond
1978–1982   Thomas F. Malone
1982–1986   Walter A. Rosenblith
1986–1990   William E. Gordon
1990–1994   James B. Wyngarden
1994–2002   F. Sherwood Rowland
2002–   Michael T. Clegg

HOME SECRETARIES

1863–1872   Wolcott Gibbs
1872–1878   Julius Erasmus Hilgard
1878–1881   J.H.C. Coffin
1881–1883   Simon Newcomb
1883–1897   Asaph Hall

1897–1901   Ira Remsen
1901–1913   Arnold Hagues
1913–1918   Arthur Louis Day
1919–1923   Charles Greeley Abbott
1923–1931   David White
1931–1951   Frederick Eugene Wright
1951–1955   Alexander Wetmore
1955–1965   Hugh Latimer Dryden
1965–1971   Merle Antony Tuve
1971–1975   Allen Varley Astin
1975–1979   David Goddard
1979–1987   Bryce Crawford
1987–1999   Peter Hamilton Raven
1999–2003   R. Stephen Berry
2003–   John I. Brauman

PRESIDENTS OF
THE NATIONAL ACADEMY OF ENGINEERING

1964–1966   Augustus B. Kinzel
1966–1970   Eric A. Walker
1970–1973   Clarence H. Linder
1973–1974   Robert C. Seamans, Jr.
1974–1975   William E. Shoupp
1975–1983   Cortland D. Perkins
1983–1995   Robert M. White
1995–1996   Harold Liebowitz
1996–   William A. Wulf

PRESIDENTS OF THE INSTITUTE OF MEDICINE

1970–1971   Robert J. Glaser (Acting)
1971–1974   John R. Hogness
1974–1975   Donald S. Frederickson
1975–1980   David A. Hamburg
1980–1985   Frederick C. Robbins
1985–1991   Samuel O. Thier
1991–1992   Stuart Bondurant (Acting)
1992–2002   Kenneth I. Shine
2002–   Harvey V. Fineberg

Notes

1. This informal review of segments of the history of the Academy has been drawn substantially from the commissioned histories by Frederick W. True and by Rexmond C. Cochrane, as well as from the memoirs of many of the individuals mentioned in it. The memoirs are written by their colleagues and published in an Academy series. Also, much of the text stems from memory, either through direct personal experiences or from extensive reading over many years. As we all know, memory can be fallible. We trust that any deviations from the truth are minor and in any case not very serious.

2. The main headings serve to introduce biographical material and accounts of special events relevant to the period in which the individual named served as president of the Academy, such as the term of Alexander Dallas Bache, which starts the volume.

3. The secondary headings are used to single out the special actions of an individual or give an account of some other matter that occurred during the term of service of a given president of the Academy.

4. Frederick True, ed., *A History of the First Half-Century of the National Academy of Sciences 1863–1913* (Baltimore: The Lord Baltimore Press, 1913).

5. Four series of lectures encompassing the most active areas of science were presented during the centennial celebration. Some two-dozen scientists who were at the peak of their areas of specialization were involved. President John F. Kennedy also spoke in commemoration of the occasion. The texts of most of the presentations were published in the book *The Scientific Endeavor: Centennial Celebration of the National Academy of Sciences* (New York: The Rockefeller University Press, 1965).

6. Rexmond C. Cochrane, *The National Academy of Sciences; The First Hundred Years, 1863–1963* (Washington, D.C.: National Academies Press, 1978).

7. The final report of the Air Force study led by Professor E. U. Condon was published under the title *Scientific Study of Unidentified Flying Objects* in 1968. It contains an introduction by the science writer Walter Sullivan, a member of the staff of *The New York Times*, and a profound essay written by Condon concerning the circumstances in which the scientific method can be applied to analyze a problem.

Daniel S. Gillmore, ed., *Scientific Study of Unidentified Flying Objects* (New York: Bantam Books, 1968).

8. The portions of this essay that deal with accounts of the buildings associated with the main Academy headquarters at 21st and Constitution Avenue in Washington D.C. may be regarded as supplementary to those that appear in the excellent article written by Detlev W. Bronk, "A National Focus of Science and Research," *Science* 176(1972): 376–80.

Index

Adams, Albert, x
Adrian, E. D., 57
Agassiz, Alexander, 22–25
Agassiz, Louis, 22–24, 34
Alberts, Bruce Michael, 93–96
Aldrich, Daniel G., 97
Allisson, Senator William B., 20
Annan, Secretary Kofi, 95
Arnold, W. D., 51
Arwade, Florence, x
Astin, Allen V., 61, 86
Avery, Oswald, 69

Bache, Alexander Dallas, 1, 2, 5, 45, 46
Bacon, Leonard L., 75
Baker, William O., 87
Bearn, Alexander, x
Bell, Alexander Graham, 20
Berkner, Lloyd, 64, 65, 85, 86
Billings, John S., 30
Bohr, Niels, 39
Brattain, Walter H., 62
Bronk, Detlev W., 57, 58, 60, 62, 65, 66, 68, 81, 84, 86, 90, 92, 94
Brooks, Senator Jack, 82
Brown, Harrison, 63, 94
Bush, Vannevar, 53, 54, 56

Campbell, William W., 43
Carmichael, Leonard, 86
Carter, Herbert E., 88, 89
Chadwick, James, 28, 29
Chittenden, R. H., 59
Churchill, Winston, 54
Cicerone, Ralph, 96–98
Clark, John C., 41
Cleveland, President Grover, 24
Cochrane, Rexmond C., 66
Coleman, John, S., ix, 86
Compton, Karl, T., 56
Conant, James B., 54, 55, 60
Condon, Edward U., 77, 78
Coolidge, President Calvin, 37
Copernicus, Nicolas, 90
Cornell, S. D., 79

Daddario, Emilio, 81, 82, 84–87
Darwin, Sir Charles, 16, 17, 24
David, Edward, 67
Davidson, George, 3, 45–47
De Forest, Lee, 51
De Gaulle, President Charles, 77, 78
Dobson, Gordon, 65
Doolittle, James H., 86
Douglas, Donald W., Jr., 86
Dubos, René, 81

Index

Dryden, Hugh L., 83, 84
DuBridge, Lee A., 67

Edison, Thomas A., 18,
Einstein, Albert, x, 25, 37, 38, 40, 90
Eisenhower, President Dwight, 55, 67

Fessenden, Reginald A., 18, 35
Flexner, Simon S., 30
Florey, Howard, 59
Franklin, Benjamin, 1

Ghirardelli, Domingo, 44
Gibbs, Josiah Willard, 7
Gibbs, Oliver Walcott, 20, 21, 24
Gilman, Daniel Coit, 26
Glennan, T. Keith, 84
Goodhue, Bertram G., 36
Gould, Benjamin, 45, 48
Grant, General and President Ulysses S.,
    7, 8, 13, 39, 41
Groves, General Leslie R., 52, 53
Guggenheim, Harry F., 86,

Haggerty, Patrick E., 62, 63, 81
Hale, George Ellery, 27, 28, 32, 34, 35
Hall, James, 34
Handler, Philip, 79, 81, 88, 89
Harkins, William D., 28
Harrison, Wallace K., 37, 66, 87
Helmholtz, Hermann, 7, 40
Henry, Joseph, ix, 3, 4, 5, 6, 7, 8, 9, 45,
    92
Hertz, Heinrich, 18
Hess, Harry, 70, 71, 72
Hitler, Adolph, 55
Hogness, John, 81
Hollmann, Hans, 64
Holmes, Arthur, 70
Hornig, Donald, 68
Humphrey, Senator Hubert, 73, 74
Huxley, Thomas, 16

Jewett, Frank Barton, 38, 49, 50–52, 55,
    57, 59

Johnson, Eldridge Reeves, 57
Johnson, President L. B., 75
Johnston, S. Paul, 86

Kelly, Mervin, 61, 62, 84, 85
Killian, James, 68
King, Clarence, 14
Kinzel, Augustus B., 86
Kistiakowsky, George B., 68
Kueter, Jeff, x

Land, Admiral Emory S., 86
Langley, Samuel P., 33
Langmuir, Irving, 50, 51
Lassman, Thomas, ix
Lawrence, Ernest O., 74
Lick, James, x, 44, 45
Lillie, Frank Rattray, 48, 49
Lincoln, President Abraham, 1, 5
Lindsay, Heather, x
Loening, Grover, 86
Lyell, Sir Charles, 16
Lyman, Theodore, 20

MacLeod, Colin, 69
Marsh, Othniel Charles, 9, 15, 16, 17,
    21
Marshall, Lauriston C., 65, 66
Marshall, Lucie, x
Mathews, D.H., 71
Maxwell, James Clerk, 18
McCarty, Maclyn, 70
McDermott, Walsh, 80, 81
McNamara, Secretary Robert, 76
Meade, General George, 20
Mellon, Paul, 87
Mendel, Gregor, 42
Michelson, Albert A., 36, 37, 39, 40, 41
Millikan, Robert A., 36, 49, 50
Moore, Gordon E., 81
Moberg, Carol, x
Morgan, Thomas Hunt, 41
Morley, Edward W., 40, 41
Mosher, Charles, 81
Muir, John, 22

Newman, Paul A., 65
Nichols, Rodney W., x, 76
Nixon, President Richard, 68
Noguchi, Hideyo, 31

Oppenheimer, J. Robert, 55, 63

Parsons, Captain William S., 83, 84
Peabody, George, 15
Pickering, William H., 67
Pinchot, Governor Gifford, 22
Piore, Emanuel, 75
Powell, John Wesley, 11, 12, 13, 34, 80
Press, Frank, 89, 90, 91, 92

Rabi, Isodore I., 51
Ramsey, Norman, 74
Remsen, Ira, 25, 26, 27
Revelle, Roger, 70, 72
Richards, Alfred Newton, 58, 59
Rockefeller, David, 63
Rockefeller, John D., 30
Rogers, Henry, 10
Rogers, William Barton, 9–11
Roosevelt, Franklin, 53
Roosevelt, President Theodore, 22
Rothenberg, Marc, ix
Rowland, Henry, x, 7, 19
Rutherford, Ernest, 28

Seaborg, Glenn T., 74, 75
Seitz, Frederick, 31, 69, 74, 76, 77, 78, 79, 86, 87
Shannon, James A., 79, 80, 87

Shockley, William S., 62
Smithson, James, 3

Thompson, Sir Benjamin, 21
Tolman, Richard, 52
True, Frederick W., 27
Truman, President Harry, 55, 56, 62
Tuve, Merle A., 83, 86

Van Allen, James A., 67, 84
Varney, Robert N., ix
Vine, F. J., 71
von Braun, Wernher, 67
von Humboldt, Alexander, 23
von Neuman, John, 21

Wolcott, 92
Wolcott, Charles Doolittle, 33, 34, 39, 92
Wall, Frederick, x
Warner, J. C., 75
Weart, Spencer, ix
Weaver, Warren, 38
Webb, James E., 66, 84, 85, 86, 87, 88
Wegener, Alfred, 70
Weinberg, Alvin M., 91
Welch, William Henry, 29, 30, 33
Wiesner, Jerome, 68
Wilson, President Woodrow, 32, 33
Wilson, Senator Henry, 1
Wilson, Tuzo, 71
Woodward, R. J., 27
Wright Brothers, 33

Yang, Quinhong, x

Figure 52.   John Wesley Powell in conversation with a native Paiute somewhere in the Western Great Basin.